THE REAL WORLD
AND MATHEMATICS

Hugh Burkhardt

The Shell Centre for Mathematical Education

D1145515

Blackie

THE REAL WORLD AND MATHEMATICS

ISBN 0 216 91084 6

First published 1981

Published by Blackie and Son Limited
Bishopbriggs, Glasgow G64 2NZ.
Furnival House, 14–18 High Holborn, London WC1V 6BX.

Filmset and printed in Great Britain by Thomson Litho Ltd, East Kilbride.

FOREWORD

This book is about learning to use mathematics. There is little doubt that the large share of school time given to mathematics reflects the long-held belief that it provides skills that are important in everyday life and work. It is no longer obvious, however, that the skills learned in secondary school mathematics *are* being much used by most people, despite a tremendous increase in the sophistication of society which has led to an explosion in the use of mathematics by a few at a high level in science and social science, technology, industry and government. Such applications as there are in school mathematics books, or examinations, mostly refer to highly stylized, artificial situations of little concern to the pupils who need much more help than this if they are effectively to use their mathematics in the real world.

In the first three chapters I aim to show that fairly simple mathematics provides a tool-kit which, when properly guided by commonsense and training, can be a powerful aid in tackling real problem situations of concern to the pupil or future adult—how to travel to school, helping in the house, playing football better, helping handicapped children, are some examples. In Chapter 1 I briefly outline the approach and discuss the essential elements of real problem solving. The emphasis is on the pupil acquiring an active attitude to new problems that arise in his world, learning to ask the questions as well as to answer them, and choosing between possible lines of attack. Since these are demanding high-level skills, the problems that are tackled will tend to be simple ones; nonetheless they should be important to the children involved. I aim to show in each case that such analysis can have a pay-off both in better understanding of what is going on and, preferably, in better decisions; the reader will probably form tentative views on this on the basis of the various situations analysed in Chapter 2, but judgement will finally rest on what pupils achieve in the classroom and outside.

The problems of teaching these skills in more detail are discussed in Chapter 3. In the last decade or so some exciting and effective work has been done and I shall describe those projects from around the world that seem to have been successful, and discuss how we can all learn to do better. There are extra demands on the teacher similar to those arising in any open-ended exploratory work by pupils—a willingness to follow and respond as well as to lead and explain, and the confidence not to need to know all the answers. However, realistic situations are easier to tackle than purely mathematical topics in that here 'commonsense' provides essential and helpful guidance, and because there are no right answers that must be found but only some answers which are better than others. It follows, however, that some knowledge of the problem situation is an essential element, for the pupil at least; this is the second reason for concentrating on problems from everyday life.

Chapters 1–3 are thus concerned with ideas, providing suggestions for those teachers who are willing to try something new and see a need to be met in helping children use mathematics. Many of these ideas are still partly speculative but the steadily increasing amount of practical experience reported here provides a reasoned basis for the approach. Most teachers will prefer to await the detailed development of this element of the curriculum over the next decade or so before considering any substantial modifications in what they do; I hope that for them too this book will have curiosity, if not Action*, value for the ideas it contains and the glimpse they present of a possible aspect of future mathematics teaching. They may even be tempted to try the occasional session or two with a class. The reader who would like to get a quick impression of the flavour of this approach, before following the coherent development we have attempted, should look at Section 1.1 (pp. 2–7), Section 1.3 (pp. 15–25), Section 2.1 (pp. 28–35), Section 2.2 (pp. 45–6), Section 2.3 (pp. 59–66), Section 2.4 (pp. 68–9) and Section 3.1 (pp. 88–9), with a passing glance at Section 1.2 (pp. 10–14).

Teachers can also use realistic situations in the straightforward, more or less didactic, teaching of mathematical applications—for example, it is surely more useful to discuss borrowing money than the traditional 'stocks and shares'. Appendix B gives a fully developed example of such a teaching unit—on kinematics in the context of traffic—and many of the examples in Chapter 2 could be taught in this way. However, such an approach does little to develop the child's ability to actively tackle the *new* problem situations that he will face.

Chapter 4 is about the tool-kit itself—it is a complementary discussion of mathematical techniques in the context of applications. After considering the somewhat different view of mathematics that is suggested by its application, I mention some important special areas and give brief reviews of those in which rapid developments are currently taking place.

I should emphasize that this work is not envisaged as a whole new approach to the teaching of mathematics but rather as an important

* See page 8.

additional activity, which might perhaps take 20 per cent of school mathematics time, justified both by its inherent value and by the reciprocal benefits it will bring to the rest of the mathematics curriculum through improved motivation, extra practice and better conceptual understanding through 'concrete' illustration. However, learning to use mathematics is the key objective—I regard *that* as an essential aspect of the secondary school curriculum which has been neglected.

Many people have contributed to the ideas presented in this book, by their suggestions and criticisms and, above all, by testing them in practice. I am grateful to John Baker, Alan Bell, Christine Bennett, Brian Bolt, Vicki Bruce, Liz Bryer, Jack Cohen, Ellen Fraser, Gwyn Gardiner, Tony Gardiner, John Godwood, Susie Groves, George Hall, John Hersee, Geoffrey Howson, Peter Holmes, Claude Janvier, David Johnson, Mary Grace Kantowski, Bob Lindsay, Earle Lomon, Brian Low, Len Masterman, Graham McCauley, Alistair McIntosh, Paul Morby, Gillian Moore, Jill Morris, Christopher Ormell, Oliver Penrose, Henry Pollak, Bill Slater, Kaye Stacey, Mike Thorpe, Paddy Turpitt, David Wishart and Ted Wragg for their help; particular thanks are due to
Sheila Bryant
Diana Burkhardt
Trevor Fletcher
Rosemary Fraser
Frank Knowles
Vern Treilibs

for substantial contributions to this work and for this reason I use the editorial 'we' in the chapters that follow.

The following organizations gave permission to reproduce copyright material, their assistance is gratefully acknowledged: Granada Publishing Ltd for the diagram on page 53, taken from *Machines, Mechanisms and Mathematics* (Mathematics for the Majority series) a Schools Council publication; Moore Publishing Company, Durham, North Carolina 27705, USA, for extracts from *Getting There* USMES Teacher Resource Book (pages 80–5); Heinemann Educational Books for the quotation from the Teacher's Guide for the *Mathematics Applicable* series, a Schools Council publication (pages 119–121); Cambridge University Press for the diagram on page 128 taken from *The School Mathematics Project Book 4*; W. Foulsham and Co., Ltd., for extracts from the *Statistics in Your World* series (The Schools Council Project on Statistical Education) reproduced on pages 143–9; the Teacher Publishing Company Ltd for the extract on page 89; the Controller of Her Majesty's Stationery Office for data from the DES report *School Population in the 1980's* (pages 172–6). The cartoons were drawn by Malcolm Swan and the cover photograph was by Paul Morby.

Prue Glass typed and retyped the manuscript, usually against unreasonable deadlines, as well as providing protection by quietly handling all kinds of administrative problems.

This book is one tribute to her memory.

Hugh Burkhardt

CONTENTS

LEARNING TO USE OUR MATHEMATICS

1.1   Problems and models—an initial glance

A real situation

"No... I won't accept American Express."

'I get a pound pocket money on Monday morning and it just seems
to disappear.'
This is a real problem. A group of children who discuss it have plenty of
suggestions, some half-joking like:
'Get some more from your dad.'
or
'Go out with your boyfriend and let him pay.'
but there are also some serious and helpful comments including:
'Where does it go?'
'Do you get paid for jobs around the house?'
'Write down what you spend every time you buy something.'
and they decide that this last suggestion is an approach worth trying.
Several children, perhaps including the originator of the problem, resolve
to keep accounts; by the next session a week or so later they have made
further progress learning, for example, that accounts generally don't
balance, that a messy jumble of items is hard to see through, and perhaps
that money has lasted rather longer than before because just recording
spending decisions makes you question them. Further discussion leads
them to try budgeting and to the realization that this is both a help and
a burden.

We shall return to this problem several times in this chapter as we discuss the tackling of real problem situations, and the teaching of skill in it. We shall see what some children made of it. For the moment it is merely a simple example of the sort of situation we have in mind—a problem of real concern, not very well specified at first, in which there could be a pay-off from a rational analysis using quite simple mathematics. Most children will not tackle such problems actively without encouragement and training; we are suggesting that there may be a multiple benefit in providing this within the school mathematics curriculum.

Modelling

How do you tackle a real problem mathematically? Figure 1.1 is a rough picture of the essential elements. We will begin on the left, which represents the problem in the world outside mathematics, then somehow we *formulate* a more or less mathematical model which mirrors at least some aspects of the problem. We rearrange or *solve* the model and then cross back to the real world by *interpreting* the answer. In practice we have also to *validate* the model by checking if the answers describe the situation well enough. Usually it is necessary to go round the processes

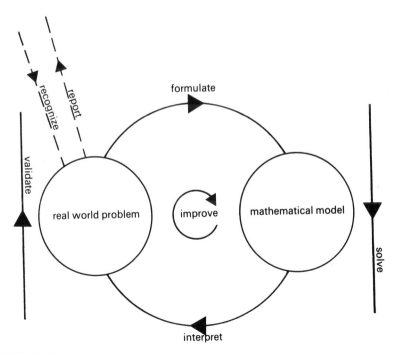

FIGURE 1.1

shown several times, *improving* or extending the model. We also have to get into and out of these modelling processes. The first involves the *recognition* that there is a problem worth thinking about, the second the communication of the conclusions including the process of clarification which exposition requires (we *report*).

We distinguish between *standard models* which describe standard situations of such general interest that it is worth teaching and learning about them, and models of *new situations* which have not been explicitly analysed before. For example,

$$s = ut + \tfrac{1}{2}gt^2$$

is a standard model for a falling body, while

'How could we get more cars through a green traffic light?'

or 'How should the goalkeeper try to block the forward's shot?'

would provide situations for analysis new to most mathematics classrooms. It is important for both teacher and pupil to be familiar with as wide a range of standard models as possible, with their uses and limitations, both because of their innate importance and because they provide examples that can be adapted for new situations. However, because time in school is limited, people's interests varied, and the world changing fast, we cannot possibly hope to teach every useful model. It is therefore important to learn the higher-level skills of model building. Because these skills are more demanding, the problems tackled must be correspondingly easier. However, it is surprising how many important problems can usefully be tackled with a very limited set of mathematical techniques, such as the abilities to:

– enumerate possibilities
– do arithmetic with a calculator
– tabulate observations and results

provided these abilities can be harnessed to the problem in hand. The abler student will, however, discover the pay-off of abstraction—for example, in the way algebra can provide, in a formula, an answer to a whole class of problems which only requires the substitution of numerical values and not the repetition of the whole argument for each case.

Illustrations and situations

There is little doubt that the large amount of time in the school curriculum devoted to mathematics is largely due to its usefulness, yet any standard mathematics textbook will be found to contain very few problems that would be recognized by the man in the street as something he would want to know the answer to. The mathematical ideas and techniques that properly form the main content of mathematics courses *are* usually illustrated with problems from outside mathematics. For example:

1 John is twice as old as Mary, $\qquad$ $J = 2M$
   who is two years younger than Ann, $\qquad$ $M = A - 2$
   John's twin sister $\qquad$ $A = J$
   from which we find the children's ages $\qquad$ $\overline{J = A = 4, M = 2}$

2 A ship leaves harbour steaming NE at 8 knots. After an hour another ship leaves steaming SE at 16 knots. How far apart are they two hours later and on what bearing?

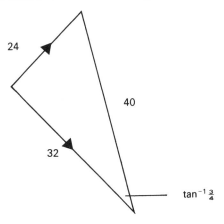

FIGURE 1.2

Figure 1.2 illustrates the answer to problem 2. Such problems have no bearing at all on any practical situation the child is likely to meet, though the second is of some interest to navigators. They are introduced to provide exercise in particular mathematical techniques—simultaneous algebraic equations or vectors.

In discussing applied mathematics, in our wide sense of the word, it is therefore important to distinguish between: *situations* by which we mean problems arising *outside* mathematics in whose understanding a variety of mathematical tools may be used, and *illustrations* which are chosen concisely to illuminate a particular mathematical point by displaying it in a concrete setting.

The dividing line is not sharp but whereas in *illustrations* mathematics (usually the topic which has just been taught) is the central interest, in *situations* it is part of the equipment for elucidating the problem, and the choice of the right tool from the available collection is part of the challenge the problem presents. Illustrations are 'clean' with neat, right answers, whereas situations usually start out with messy, not very well-defined questions which have to be cleaned up in their solution, e.g.:

'Should I go to school on my bike?'

'Should I get a record player or a cassette recorder?'

'My sister doesn't do any of the chores but I always get nagged.'
are situations raised as problems by children. You may like to think about how you might help a group of 13-year-olds to tackle them; we shall discuss them further in Chapter 2.

As with any curriculum change, new demands on the teacher are implied by some of these ideas, particularly that of tackling unpredigested problems from real life with no unique answers. The teacher will find it pays to be a moderator of discussions rather than a lecturer, to answer questions with questions, to listen as well as to talk, and generally to throw responsibility back onto the pupils much more than is usual in mathematics classrooms. While it requires confidence to begin to explore this approach, many teachers will find it a valuable way of refreshing their lessons, and helping their pupils to link mathematics with their experience of the wider world.

Illustrations, such as the two artificial problems just discussed, have an important place in developing mathematical technique and suggesting ways in which more realistic models might be built. Even so, it would be worth looking for more relevant illustrations—football, for example.

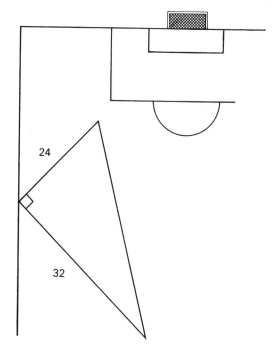

FIGURE 1.3

3 An attacking player on the left touch-line starts to run at $8\,\mathrm{m\,s^{-1}}$ and at 45° to the touch-line towards the goal; at the same

moment the mid-field player, 32 metres away at right angles to his run, passes the ball so that it just reaches the running attacker 3 seconds later. At what angle should the mid-field player aim the pass and at what horizontal speed?

This involves much the same mathematics as the 'ships' problem and the result is of interest to many more pupils; it is perhaps even surprising how far ahead of the player the pass should be aimed. However, the student needs much more help than is currently provided if he is to be able to transfer the skills he has learnt, and practised in such exercises, to the tackling of complete real problems. An important aim of mathematics teaching should be to provide this help.

Conkers

As a gentle transition to the realistic problems that form the main theme of this book let us look in some detail at a very simple nearly-real problem. We have a group of boys, say 4 of them, who go out to collect conkers. After a while they return with the conkers—there are perhaps 12 of them in the bag. How do they divide them among themselves?

The immediate response is to say, 'Ah ha, there are 12 conkers and there are 4 boys. Yes, well the answer is that each boy will get 12 divided by 4, that is, 3 conkers'. Surely we would all, having written that down and underlined it, have felt that we have provided an adequate solution to the problem from the point of view of the mathematics lesson. Do you think that 4 small boys would regard this as an adequate solution and, if not, what has been left out?

At this stage groups who have discussed this matter recognize that there is still a problem because all conkers are not the same and, by saying that each boy has 3 conkers, you have not solved the problem of distribution—which 3 conkers each boy has. So what would happen next? The boys would perhaps lay out the conkers and select one each in turn, going round until all the conkers were used, so the first boy would get the first, fifth and ninth choices, etc. Notice now that this method of procedure makes the division sum redundant—the process of distribution does the division for you automatically. So that is a second possible model of how to divide the conkers, more complete than the first one. But is it complete enough for small boys? Well, no.

There is clearly the problem of who chooses first. Here too a number of solutions are possible. The largest boy may insist on choosing first, or he may just take all the conkers. Perhaps the one who collected most may be allowed to choose first. They may 'toss' for it or draw straws. Or the cleverest boy may suggest 'eeny-meeny-miny-mo...', adjusting the starting point so that he wins—this is an apparently random procedure that is in fact deterministic. These approaches have different concepts of justice based on power, work, equal chance, or cunning respectively.

You may not be satisfied with this solution and we would support your right not to be. More important, the boys may not be satisfied—they might want unequal numbers if the conkers were very different in quality. In realistic problems there are no right answers; modelling is a process of increasing understanding with room for differences of opinion. There are, however, wrong answers—for example each boy gets 12 minus 4, that is, 8 conkers.

We shall come back to this example when we look at the problem-solving process in Section 1.2.

Levels of interest

In choosing problems for analysis the interest level for the student is obviously an important factor. The following classification may be helpful.

Action problems are those whose answers may directly affect decisions in our everyday lives.

'How can I fit in my homework with the TV and going out?'

Believable problems are those that we can recognize as Action problems either for ourselves in the future or for someone we care about.

'Should I get a job at 18 or go to college?'

Curious problems are those which intrigue us, either because the phenomenon being studied is itself intriguing or because the analysis is.

'Why are there two high tides each day?'

Dubious problems are there simply to provide exercise in mathematical technique.

See p. 5, or any mathematics exam paper.

Educational problems are a rather special category—they are essentially Dubious but make an important point of mathematical (or physical or economic) principle so clearly and beautifully that no-one would want to get rid of them.

'If I invested 1p at 5% compound interest in 512 AD, what is it worth now?'

On this scale mathematical education has rarely aimed at a higher interest level than Curiosity and a very large proportion of the problems that we ask children to do are frankly dubious from an educational point of view. Indeed one essential quality of a good teacher is the ability to make Dubious problems into Curious ones for many of his pupils, but it may be that capacity for curiosity is fairly highly correlated with intelligence, making it unsuitable as an *exclusive* basis for the mathematical diet of most children. Since mathematics itself is abstract, problems of higher interest level must be sought through its applications. In this book we are suggesting that school applied mathematics should be, at least in part, based on Action problems and Believable problems for the pupils concerned. We shall describe and illustrate some methods that are being developed that enable this to be done.

The obvious areas of high interest in the 11–16 age-range include: money, sport, music, clothes, personal relationships (including sex), school work and leisure. Other, perhaps less compelling, areas are: transport and traffic, holidays, school organization, careers planning. We shall make suggestions for work in all these areas. Though a tremendous amount remains to be done in collecting and developing material, in many of them there is published material providing standard models for classroom use. Ironically much of it was developed for less able pupils, some in response to the raising of the school leaving age. It is clear from experiments that this approach can be, as usual, of even greater benefit to the more able who can handle more of its substantial challenges.

In seeking everyday problems we give a high priority to the interest level but we also seek some sort of pay-off from very simple analysis, as an encouragement for pressing on with the challenging business of trying to probe more deeply. The world is full of such situations but it takes a practised eye and a fair amount of imagination to see them. The essential skill is to look only for a problem and not to ask if you can do anything about solving it until afterwards.

Finding problems—ask the children or PAMELA

Problems play a crucial role in learning mathematics of all sorts, since it is an inherently active subject. In conventional areas large collections of illustrative problems have been built up over the years. In any new approach, therefore, finding enough suitable problems is an important challenge. The building of a collection of rich, realistic problems, and of models useful in their understanding, is important in developing the teaching of realistic applied mathematics. PAMELA is such a collection of problems from everyday life—it is described in Section 3.1. However, one effective way of finding good problems is to ask the class.

Ask your students to write down a problem that they have been concerned about in the last day or so—the protection of anonymity is important because many of the problems will raise sensitive issues. Collect the papers and choose a hopeful-looking problem, or read out the suggestions and let the group decide what they would like to study. You are likely to find many interesting problems, and some in which the pay-off from a mathematical analysis is clear. Others may prove less profitable and thus contribute to a balanced view of what can be achieved.

1.2   The problem-solving process

Now let us examine in more detail how we tackle real problems. Figure 1.4 is a somewhat more detailed analysis of the steps in problem solving which were outlined in Figure 1.1. Keep in mind the conkers and pocket money examples as we work through the diagram; we shall relate them both to the scheme.

The left-hand part of Figure 1.4 is a conventional flowchart. This piece of mathematics is a relatively new addition to the curriculum.

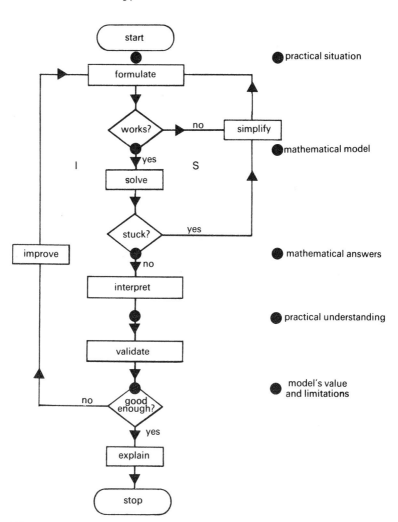

FIGURE 1.4

Though superficially it is just a list of actions and decisions enclosed in boxes, the fact that the arrows join the boxes, clearly defining a variety of possible paths, makes flowcharts much more flexible than a list of instructions written on a piece of paper. For this reason complicated government regulations are sometimes specified in flowchart form. One weakness of a flowchart is that it emphasizes actions or processes rather than the states of the system between those actions. On the right-hand side of Figure 1.4, therefore, we have listed the state that the problem-solving process has reached at the corresponding blob on the flowchart.

Starting at the top of the diagram with the practical situation, we formulate a model; we shall discuss what this involves in more detail in Chapter 3—for the moment we just 'think of something'. In flowcharts, diamonds represent points of decision, for example, 'works?' asks us to decide whether the model is complete enough to give us some answers. If the answer is 'no' we go back and formulate some more relations, otherwise we go on and solve the model mathematically giving, unless we get stuck, some explicit mathematical answers to interpret in terms of the practical situation. We validate the model by comparing our answers with our knowledge, from observation or experiment, of the situation. If the model is not good enough we attempt to improve it, going back to formulate a better model.

How do we decide whether a model is satisfactory or not? These criteria are really part of the formulation process. In formulating a model there is some purpose we have in mind for it. The model will be satisfactory if it describes the features of the situation which are essential for this purpose, and with sufficient accuracy for the questions we want to ask or the decisions we want to make. Few human decisions are disturbed by errors of a few per cent. We don't usually need to know the speed of a car, the brightness of a light, the size of a meal or the length of a lesson more accurately than this. Financial transactions do need to be more accurate; we would all notice if our pay cheques were 2 per cent wrong at the end of the week or month. But even here the value in *real terms* of the money we receive is subject to greater uncertainties than this. Few of us know by how much the real value of our salaries has gone up or down since last summer. The *stability* of the results is crucial—if a small change in the assumptions produces a large change in the conclusions, the model is less likely to be useful.

How does our modelling of the conkers problem fit this scheme? Figure 1.5 gives the states of the problem corresponding to the right-hand column of Figure 1.4. The three columns correspond to our three gradually improving models. First we concentrated on numbers only. T is the total number of conkers, b the number of boys and n is the number each boy would get. The model

$$T = bn$$

is a mathematical statement of equal division. Knowing T and b we can solve for n and interpret the answer in words. The obvious limitation leads to the second 'distribution' column and so on. The validation in each case includes seeing if the model does work.

conkers

aspects of situation	numbers	distribution	'fair order'
model	$T = bn$	choose in turn	biggest or best or random
answers	$n = \dfrac{T}{b} = \dfrac{12}{4} = 3$	$\begin{pmatrix}1\\5\\9\end{pmatrix}\begin{pmatrix}2\\6\\10\end{pmatrix}\begin{pmatrix}3\\7\\11\end{pmatrix}\begin{pmatrix}4\\8\\12\end{pmatrix}$	
understanding	3 conkers each	4 groups of three	
value and limitations	which conkers?	which boys?	OK?

T = total number of conkers
b = number of boys
n = number of conkers for each boy

FIGURE 1.5

Realistic applications are naturally and essentially linked to observation and experiment, both in the formulation stage and particularly in validating the predictions of the model. Experimental work and data collection both cost time and money and most mathematics teachers have little practice in them. Some such work is essential in studying realistic situations but experimental awareness can be economically extended by using recorded visual material from a film or videotape, or by computer simulations where available. But often the pupils' experience will give them a clear, and reasonable, view of the value of the model.

Figure 1.6 gives a similar outline of one discussion of the pocket money problem with which this chapter started. Notice again the looping structure representing successive improvement of the model.

We shall now comment in more detail on the various processes involved in modelling as illustrated in Figure 1.4.

1 The first step, *formulating* the mathematical model of the situation, is the most complex, difficult and challenging. It involves an understanding of the natural laws of the system, in identifying the important variables and finding enough relations

pocket money

aspects	where does it go?	what kind of buys?	what do I want?
model	record + add $$T = \sum_i P_i$$	group like buys $$G^{(n)} = \sum_i P_i^{(n)}$$ $$T = \sum_n G^{(n)}$$	budget
answers	$\{P_{ij}\}$	$G^{(n)}$	intended $\bar{G}^{(n)}$
understanding	list of purchases	how much spent on different kinds?	spend up to $\bar{G}^{(n)}$ on n things
limitations	doesn't add up too many to 'see through'	has not led to action	OK?

T = total pocket money
$P_i^{(n)}$ = i-th purchase (type n)
$G^{(n)}$ = total spent on type n
$\bar{G}^{(n)}$ = budget for type n

FIGURE 1.6

between them to produce some results.

In the traditional teaching of applications this step is short-circuited; the student is shown the situation and at once *given* a model to describe it. In problems, only very minor modifications (e.g. different numbers) are encountered, and exactly the right information is always given. This is a useful form of activity, designed to give some familiarity with what is known (like reading Shakespeare in English lessons). It does not help the student to face a new situation, however simple (like writing a letter)—this is our objective in teaching modelling. Although formulation is an intuitive and partly creative process, most children can make satisfactory progress in it provided the problems are simple enough and the framework is right.

2 The second step—*solving* the mathematical model—is the most familiar. It is here that nearly all the effort of mathematics teaching now goes.

3 The *interpretation* step is usually straightforward, once the formulation has been worked through. However, it is usually omitted—a neatly underlined correct answer does not mean that

the student understands the practical implications of what he has found.

4 The *validation* or *checking* step is vital. All models of real situations are inadequate in some respects and it is important to know where they work and where they don't, and to what accuracy. It is also important to check the mathematics (step 2)—everybody makes mistakes. This validation process is also usually ignored.

These then are the essential steps in solving a problem, but it is most unusual to go straight through them and be satisfied. The two iterative 'loops' usually involved are shown in Figure 1.4. They are related to two phases in solving a problem:

(S) the simplification phase

(I) the improvement phase.

When you start to look at a real practical situation you are confused by its complexity and unsure what to tackle first. The first formulation step (1) consists of *identifying the relevant variables* and *some simple key questions* to ask about them, before going on to try to answer them by *formulating relations* between them. It is a process of cutting down the full problem to something tractable. The first ideas for a model may still be too difficult to be solved within the limitations of time and energy available; if this is so, we must resort to a further simplification indicated by the inner loop S in Figure 1.4. Later, when an initial simple question has been answered, you normally build on it, elaborating the model to answer more questions or to give more accurate answers—this is the improvement phase (shown by the outer loop in the figure). Finally you stop either because the model is adequate for your objective, or from frustration or fatigue.

All these phases are illustrated by the problems of Figures 1.5 and 1.6. In each case a start is made with a simple model for part of the problem, which is clearly inadequate. This is later improved. In each case more could be done, and there are alternative routes through the problem that could be taken, but useful insights have emerged. In each case mathematics provides only one part, but an essential part, of the equipment used; the range of modelling skills is much wider and involves a lot of commonsense.

In this section we have presented a model of modelling. It too has its limitations—it is undoubtedly an oversimplification, presenting just one picture of the complex thought processes involved in tackling real problems. It should not be taken too seriously but may be helpful in giving us some awareness of the nature of these skills. We take it further in Section 3.3. Here and in the main part of the book we shall largely concentrate on problems and mathematical techniques relevant to the 11–16 age range. However, in order to show that the principles are quite general we discuss an Action problem in education in Appendix A—the supply of teachers.

1.3 Teaching mathematical modelling*

Real situations in the classroom

By now we hope that you recognize modelling as a fairly familiar activity—like Molière's Monsieur Jourdain, who was surprised to discover that he had been talking prose all his life, we have all been tackling real problems more-or-less coherently in this way for a long time. We conclude this chapter by discussing briefly how one can teach these skills in the school classroom, making explicit suggestions for teachers to experiment with, and reporting what some others have found. Chapter 3 gives a fuller discussion of problems of teaching and assessment.

There are no widely established methods of teaching the skills of modelling real situations to children. However, some things do seem to be important. Only mathematics that is very well absorbed seems to be usable in modelling. The problem situation must be simple if the pupil is to successfully develop models for himself—far simpler than when he has the model provided for him by the teacher. The teacher should provide the minimum support needed to avoid discouragement—we believe that, generally, it pays for him to know the problem fairly well. The initial simplification phase (cutting the problem down to a reasonably well-

* This section was written with Rosemary Fraser.

defined simple question with a provisional way of answering it) takes a substantial fraction of the time that it is reasonable to spend on the problem. You may need to encourage the creative generation of ideas. Generally, pupils have learnt above all to avoid making incorrect remarks; explain that the ideas can be sorted out and improved afterwards and you will usually get a fair flow. The initial discussion is often apparently random and disconnected; the teacher may need to help in listing and structuring the suggestions. Formulation seems to be much easier in groups than for individuals working alone. Assessment of the more creative aspects of modelling skills is difficult and probably unnecessary; to demand a written account, or an oral presentation, of the problem and its understanding, develops the communication skills which are so essential to the 'applied mathematician'—a concise report, in note form with diagrams and tables, is better than a diffuse essay.

One way to start

It may be helpful to make specific detailed suggestions as to how an initial session might be run. What follows is an approach that has been found successful but it will surely not be the best way for everyone. The teacher, as he acquires experience, will vary the approach looking for a style which suits him. These suggestions are explicitly related to the first few sessions—once the class has some experience, the pupils will take over more of the suggested activities.

Choose a double period so that at least one hour, and preferably more, is available. Make it clear to the class that you are going to try something new—looking at real problems. It should be reasonably good fun but it is serious, though it may not look like mathematics. Mention a particular problem as an example—perhaps pocket money budgeting or travel to school.

Ask the class to suggest problems, as described in Section 1.1, page 9. If there is a suitable problem use it—otherwise discuss some of them with the class but return to the one you've already mentioned for detailed study. Try to come back to one of their problems on another occasion at least, as this will increase the pupils' sense of involvement and responsibility for the problem and its solution. State the essence of the problem very briefly then say, 'It's over to you!'

Ask for any ideas—wild ones too. List them on the blackboard or the OHP, putting them in columns with the more tractable-looking ideas in the middle, and the vaguer ones at the right. When things start slowing down suggest that they pick out *variables* (quantities that you may need to discuss) and list these at the left, giving them algebraic letter names 'for shorthand'. This opening phase usually flows easily—it could well take twenty minutes or more and may, for example, produce a

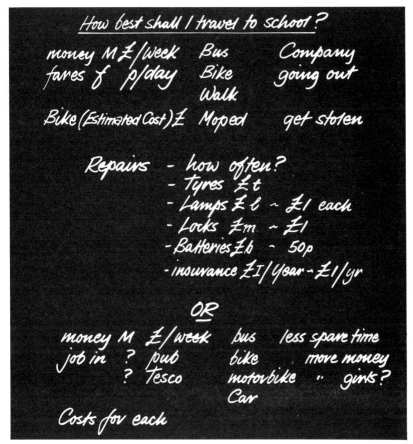

FIGURE 1.7

blackboard looking like Figure 1.7 for the problem 'How best shall I travel to school?'

The first crunch point has arrived. We have a whole list of things to talk about and now we need an actual *relation*, a statement that says something. It pays to look for something very simple that sounds as though it might be sensible. If it is too simple or not quite sensible it can be improved; the important thing is to make a start. In this case you will probably get something directly useful.

'Let's work out how much money we get for travelling to school.'

'I get £1.50 a week for the bus.'

'How much is that a year?'

'£1.50 × 50 weeks is £75.'

'We are not at school every week.'

And they're off. The group can now choose which approach to follow and start working it through, 'getting numbers' and working things out.

At this point it is worth breaking up the class into groups of 3 to 6 pupils, so that everybody can take an active role—each group might work on a different mode of travel, for instance, though it can also be useful to have different groups working on the same problem at a time. Later the groups can report and compare notes; when they do, the validation of their arguments will come from each other's criticisms as well as from suggestions for getting more objective data. Questions from the other groups—e.g. arguments about numbers assumed—may lead to a discussion as to how and where they can be found more reliably, and so on. Don't attempt to wrap it all up at the first session. Explain that each group should follow up its ideas and tidy them up for presentation at the next session, which should be at least a week later to allow time for thoughts to mature and for the energetic to gather information. Criticism and ideas at the second session, which might be fitted in alongside a start on a new problem, should often lead to a draft report and a third look.

The *report* is an important element in learning to model. It should be in the form of 'advice to a friend' on the problem situation—this requires clear explanations, which the pupils may justly claim are not necessary for their own understanding. This has value in several ways— communication skills are useful and important in themselves, while the forced 'externalization' of the argument makes the student more aware of the need to improve it, and also gives the teacher a much clearer view of progress which is valuable for diagnosis of pupils' difficulties. For all these reasons it is often good to ask every pupil to write either a separate report or a specific part of a joint report.

It is important to recognize that the conclusion of the analysis may be unsound, and that the judgement as to how far to accept it will be a largely intuitive decision for the group to take. The history of government inquiries and other forms of long-range planning, in particular, shows that even careful model building can fail to balance all the factors involved in an acceptable way—the training of teachers (see Appendix A, page 169) and doctors and the planning of the third London airport provide fairly recent examples. However, even here we believe that the intuitive decision is likely to be sounder in the light of the insight the model provides; in very many other less complex situations the model's conclusions will be directly and clearly of value.

Finally, we should often aim not only for better understanding but for positive action. If on the basis of what has been learnt it seems that something could be done better, it is worth trying to get it done better, and to check later that the change has helped. Often this may only involve individual pupils but sometimes the class, the school or other authorities may need to be approached. Occasional successes of this kind are very good for morale (see the USMES project in Section 3.1).

We repeat that the above are only suggestions—many variations are possible. For example, although the opening 'ideas' phase can work well with a class of 30, when they divide into small groups the teacher may be hard pressed in getting them all started on more detailed investigations at the same time. This can be obviated by initially giving most of the groups some more familiar mathematical activity to pursue, while starting off the others on a realistic problem. This tends to be a difficulty only in the first session or two; though group activity always demands careful organization, *that* gradually becomes part of the group responsibility for the problem. As the pupils become more experienced, of course, they become better at handling the initial phases themselves.

The report that follows describes one teacher's reaction to her first attempts, with some examples of what the children produced.

Problem solving with 1E

'A class of 30 middle-to-low ability 11- and 12-year olds, on being asked to record their current problems, reacted with a recognizable sense of apprehension at the sudden shift of responsibility from the teacher to themselves. To allay this fear the next 5 minutes were spent organizing the class into groups—this allowed time for thoughts to collect without stress. It was interesting to read later that several children's problems indicated a fear of underperforming at critical times, particularly in the answering of direct questions.

My own fear was that of an unfamiliar role in the classroom and what I should do if *no* problems appeared on the children's papers. From the response of this class it seems that this is an unnecessary fear—there was no lack of problems. Once the children started, most listed more than one problem and some children made rather lengthy lists—maybe I should have asked for their most important problem!

(DJ) keeps pricking me with a sharp pencil

I only get £1.00 Pocket money

I have a goldfish called Jaws and every time I get another goldfish Jaws eats them

Concern about homework formed the common thread—for some children it appeared to be an ongoing worry as they seemed unable to retain comprehension and were often in trouble with their work. Others were expressing the common grouse against authorities that dictate.

My problem is I can't do my homework because I keep on watching television.

Homework should be the discussion point, I decided, and spent the next 15 minutes drawing in ideas. With 30 children of this age and ability, this is not an easy task, and I felt my attempt was unproductive, although we finished with some listed questions and the groups returned to work using these as a basis.

1/ class works to easy

2/ work should be 10 times harder

3/ People who do not pass should not have to pay for exams when people who did do not pay.

4/ Dinner hours to long.

They found it difficult to record their thoughts clearly and obviously needed help to achieve this. Nobody made a neat list of time spent on homework or added up to find the total hours. Thus not even addition was being applied. Discussion ranged on regarding problems and I felt that we were developing a 'grumble' session instead of applying mathematics.

The children had also mentioned pocket money and I asked them to record how they spent it during the next 8 days. We discovered that the amount of pocket money received ranged from 35p (5p a day) to £3 with the mode at £1.

My feeling at the end of this 80 minute session was that I had most certainly learnt a tremendous amount about the children's personalities. We had not done any significant mathematics, but already topics that would form useful examples for formal lessons were building up in my mind—even,

'My hands shrink when I wash up.'

might be worth discussing when looking at enlargements! However, could such lessons continue on the basis of perhaps one a fortnight, and contribute eventually to the development of the children's ability to apply mathematics? There is obviously a need for the children in this class to begin to learn how to record, and perhaps list items.

They brought information regarding their pocket money to the second session. This was most often in the form of a chatty paragraph and contained very little in the way of detailed breakdown of cost of items, etc. On this occasion I avoided the full class discussion and each group was given large sheets on which to record information. In this way

POCKET MONEY

SOMETIMES I GET 75P A WEEK OR I HAVE IT DURING THE WEEK. I SPEND MY MONEY ON 15P ICE CREAM I GO TO THE DISCO WHICH COSTS 40P TO GET IN AND THE MONEY LEFT I SAVE IT. IT COMES TO 20P. MY MUM GIVES ME MY BUS FARE. THE THINGS I WANT TO BUY WITH MY MONEY IS EARINGS AND CLOTHES.

Pocket Money

I have £ 1.00 a week

10p for sweets in the morning

15p for ice cream in the evening

50p for films on Saturdays

10p for bus to go to films

15p on sweets in the films

100p a week

motivation was improved—the large felt pens proved very popular. As prices had not on the whole been noted and also because not many of the children assumed responsibility for their spending, I made a suggestion that items that they desired to buy per week be listed and then during the week the prices added. Some children then asked if they might do Christmas lists, while others became interested in the cost of

CHRISTMAS LIST

NAME	PRESENTS	PRICE
MUM	PERFUME	£ 2.00
DAD	AFTERSHAVE	£ 2.00
CAROLINE	RECORD	£ 1.00
MICHELLE	SELECTION BOX	95P
NAN	SOAP	75P
GRANDAD	CIGARS	75P
TIGER (CAT)	TOY MOUSE	25P

clothes. We now had plenty of possible work for this session and to feed into other maths lessons (not just for this group but for other maths groups as well). Although pupils can only model with mathematics they have fully absorbed, the study of real situations can on occasions throw up the need for a new mathematical technique which can be a strong motivatory force for learning it, and thoroughly!

The atmosphere became more relaxed as both I and the children grew accustomed to the freer situation. I felt encouraged to continue this type of work which I would describe not only as 'problem solving' but as 'responsibility sharing'. A long-term plan would be to use one lesson a fortnight, encouraging the children's co-operation—they would be able to bring formal problems or we might tackle some general problems that others suggested. Placing the children in a consultative role on others' problems does seem to encourage response. Perhaps on one occasion we might tackle a *fun* problem, but the important aspect to my mind would be presenting the children with an opportunity to express themselves and to work collectively on problems of concern to them.

Of course, older or more able children can quickly achieve more decisive results.' (Figure 1.14).

Some comments

The experience reported in the previous section was valuable for the teacher because of what was learnt about the children's personalities, abilities and interest and because it supplied information and topics that could be used in later, perhaps more traditional, lessons. But was it valuable to the pupils and was it really applying mathematics? We contend that for an initial session with a class of this age and ability it was both of these, even though the only skills that were obviously applied were the mathematical skill of addition and the more general problem-solving skill of recording information systematically. It is a sobering experience for a teacher to discover that even these skills are difficult for pupils to apply independently, but it highlights the need for pupils to be given some training and practice in using what mathematics they know to help understand their own problem situations.

Some people feel that tackling new problem situations is research, which in a sense it is, and that it is unreasonable to expect any but the extremely able even to attempt it. There is enough experience around the world to show that this is much too pessimistic. The achievements of even quite young children on suitably challenging situations are remarkable. The USMES project in the United States (see page 80) has shown that classes of primary school children, working together over a period of weeks, can gain substantial understanding of quite complex situations involving science, mathematics and human interactions. We do *not* mean to suggest that the modelling of new unstructured problem situations is a complete balanced diet in applied mathematics— familiarity with a range of well-established models remains essential— but it is a vital ingredient in training people to use their mathematics.

We have stressed the building of *mathematical* models because that is the skill we are seeking to develop. They are, however, only a very

Ski-Trip /78-79

Money Problem for clothing and pocket money.

Clothes needed:

		Cost
1) Anorack		£9·00
3) trousers		£18·00
1) goggles		£1·00
1) waterproof dungarees		£10·00
1) pair gloves		got 'em
2) tights		£3·00
7) socks		got 'em
2) big socks		£3·00
3) Jumpers		£10·00
1) hat		got 'em
1) scarf		got 'em
7) Pants		got 'em
3) shirts		
1) indoor shoes		got 'em
1) walking boots		got 'em
		£55·00

£85 to get between Nov 5th — April 5th
£20 pocket money for trip
£55 money for clothes

Nov = 25 days
Dec = 31
Jan = 31
Feb = 28
March-31
April= 5
181 days

21 weeks to get £55· £1·50 pocket money per week   21 weeks

£1·00 saved a week for 21 weeks = £21
                          Christmas = £10
                          Birthday = £10
                                      £41
total cost parents pay for extras =   -41
                                      £34
                                       75

FIGURE 1.14

small part of useful modelling (they are useful because they are powerful and cheap). In the wider sense, there are of course physical models of all sorts; apart from their own importance it is often a physical model which leads us to a mathematical formulation. For example, a clock is rarely as simple as the idealized compound pendulum we use in our models. Similarly, mathematical models of river pollution, which use differential equations to describe the dispersal of pollutants which enter at various sources along the river, have been used successfully in planning the restoration of many rivers including the Thames; equally, small-scale physical models of river systems have been constructed and have proved valuable for the same purpose. Simple physical models may be worth using in teaching modelling—in traffic problems for example, or in financial transactions with younger children. For those with the resources, computer simulations are a powerful way of building understanding—they form a useful conceptual bridge between physical, numerical and more abstract models. (See Section 4.3, page 130.)

The practical advantages of a model, physical or mathematical, lie in many directions. We have stressed the understanding that can be gained with relatively little effort. In industry, government or research it is often the cost that is crucial—mathematics is relatively inexpensive.

In some situations the motivation is even stronger. 'A terrorist plans to drop a percussion grenade out of the window of his getaway car, reasoning that when it hits the ground his pursuer will be right on top of it.' A journey to the moon, or a coronation, are other situations where the opportunities for experiment are limited. Action based on such unvalidated models is clearly more hazardous.

We have so far used modelling to describe the construction of quite explicit mathematical or physical models of problem situations. Modelling, however, is a much more general mental activity than this; indeed it can be maintained that all thinking is modelling in that, using language and pictures, we construct in our mind a more or less coherent picture of the world, which is certainly a model. However, since anything so universal begins to lack a specific identity perhaps we should leave philosophy and return to mathematics.

2

SITUATIONS IN EVERYDAY LIFE

The power of mathematics shows through its applications in almost every field of human activity. The array of models that have been produced and the phenomena they describe are a major human achievement and the serious student of mathematics will want to know a good deal about some of the most impressive of them—classical Newtonian mechanics, quantum theoretical physics, the theory of the control of systems in engineering, biology and economics, electronics and information theory and so on. However, an appreciation of these eminences requires a knowledge of their empirical backgrounds as well as a power in mathematics well beyond most pupils of 16—a similar argument can be applied to the analogous study of Shakespeare's plays (or Homer's *Iliad*) though it is probably less compelling. Nonetheless, there are standard models of power, and usually of elegance too, which are accessible; some of these are familiar.

In this book we shall largely confine ourselves to everyday problems which are recognizable as such to the mythical man in the street. There are a number of important advantages in using these situations as the basis for applying mathematics.

1 Pupils are familiar with the situation and have a firmer empirical basis for choosing appropriate models and evaluating their predictions sensibly; this leaves more of their attention for the challenge of understanding the models themselves. In more traditional areas such as mechanics or statistics the phenomena themselves are difficult and often ill-understood.

2 Following from this, less time need be spent in teaching the empirical background, which must be well assimilated if the student is to take an active role in modelling it.

3 Motivation can be high if there seems to be some pay-off from analysing the situation in equipping the student better to cope with it. It is not necessary that every problem tackled should be an Action problem for every student and show a direct pay-off (indeed provoking curiosity is a major objective), but action should emerge as a central feature of applying mathematics.

4 Students will gradually come to look at the world with an active analytical eye, learning to *ask* questions in the expectation of finding useful answers.

The problems we shall discuss in this chapter have been chosen to show how a variety of aspects of life may be analysed at a range of technical levels. Some of the situations are obvious and mundane while others are, we hope, less conventional but still interesting; in almost all of them we see some potential pay-off in the greater awareness that the analysis provides. We assume as a basis that the pupil can use a calculator, and can collect data and represent it graphically. However, we also aim to show that much more mathematical power than this pays off. Even in quite simple economic problems, algebra is a real boon either because you can work out an expression for the answer which covers a range of situations, or because the problem is more easily formulated *implicitly*, giving algebraic equations to solve.

At the other extreme, some of the models we present are almost entirely qualitative, involving a rational analysis of the situation with little or no explicit mathematics—Section 2.4 on human relationships contains several examples. We believe such discussions can be helpful. The ability to think sensibly, particularly about emotionally charged situations, is valuable, while the key elements of any decision often rest upon qualitative features of the problem. Although, when possible, quantitative discussion makes things clearer and surer, it is the structural analysis that is crucial. Mathematical tools such as graphs and tables may promote clear thinking and organize decision making.

Most of the descriptions that we give in this chapter are brief outlines designed to show some of the important elements that might emerge in a group attempt to tackle a problem, though in some cases there is a fuller discussion. We are not, of course, presenting 'right answers', and all the models need to be examined critically. We hope that in many cases the teacher will feel that he, and his class, could do better.

2.1 Money and time

Money is probably the commodity in our lives that most demands quantitative thinking. Though it does not loom very large for the young

child (who will practise counting breaths, fingers, sweets, people and so on, and will begin to measure things against his height or reach) by the age of 11 the essential facts of economics are familiar in the limited universe of pocket money, birthday presents from unimaginative (or understanding) relatives, and school dinner money. Not only can most children pay and receive change with precision but they understand that money provides a framework in which unlike things are related; you *can*, in this sense, add two bars of chocolate to a quarter of a pound of jelly babies—you get 45p (readers will probably need to multiply this by about 1.1 for each year since 1978, illustrating another vital economic phenomenon). Although most of their needs are still provided in kind, children are aware of the constraints that limited money imposes on their decisions; the shortages of the rich—time and energy and imagination— are not yet so pressing.

Money is *the* quantifiable variable in most of our lives. This alone justifies a prominent place for it in the school mathematics syllabus; however, there are further compelling reasons. As many of the phenomena associated with money are already familiar to the 11-year old, so the exacting demands of model building lie only in appreciating the properties of the mathematics. Furthermore, motivation is potentially strong. Crucially, economics is not taught explicitly in the lower school so the mathematics teacher must try to fill this role as well. Although some may be unhappy to accept this burden, as lying outside their professional competence, many know the benefits to mathematical understanding that flow from familiarity with an important application.

Ian's pocket money

Ian went home from school one day with an almost legible list of the pocket money each child in his class received each week. (Figure 2.1)

Name	Amount in P	Name	Amount in P	Name	Amount in P
Tom	75	Sandra	100	Sara	100
Mark	100	Jacky	200	Martha	100
Philip	75	Jane	150	Ian	75
Juliet	125	Broose	150	Gillian	220
Dean	200	Lauren	75	Amanda	150
Ian	125	Noel	200	Joan	100
Paul	225	Paul	100	Philipa	100
Richard	100	Luiji	100	Robert	125
Emma	100	Karen	125	Margret	100
Kathy	250	Sharon	125	Antony	125

FIGURE 2.1

He had even added a histogram to reinforce his point. (Figure 2.2)

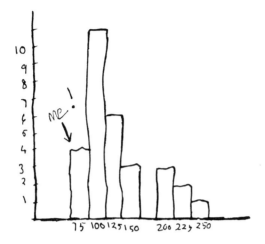

FIGURE 2.2

He resisted his parents' suggestion that the figures were not comparable—'Are you sure the others don't have to buy dinners out of theirs?'—and persisted through some passive resistance. In the end the parents retreated in fairly good order and he carried his elder brother and sister along with him to victory and, apparently, £1.00 per week each. In view of this, it is carping to point out that his case might have been even stronger if he had used a *cumulative* frequency distribution. (Figure 2.3)

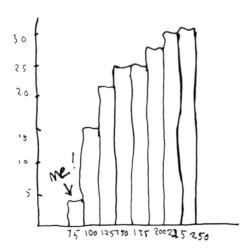

FIGURE 2.3

Travel to school

As children, particularly boys, enter their teens they usually become interested in personal means of transport—bikes, mopeds, motorbikes and cars move successively through the centre of their field of interest as their financial resources develop. Money, however, remains a recurrent problem. A discussion of using an alternative mode of travel to school has a firm base, since most parents provide money for the bus. It might be

$$\text{Bus} \approx £1.50 \text{ per week} = £60 \text{ per school year}$$

which defines a standard of comparison.

A bicycle is an obvious alternative—many boys will have one already but may well have ambitions for something lighter, faster and otherwise more exotic. How much does a bike cost? £60 is the *capital* cost of a pretty basic bike. How can that be linked to a *cost per year*? You will get suggestions based on borrowing and paying back the money—or rather the extra money that is still needed after savings, expanded birthday and Christmas presents and the profits from the second-hand sale of the old bike have been taken into account. These costs could be linked to the formal borrowing costs of hire purchase, see page 35.

What other costs are there? The repairs on a basic bike need not be expensive; they are hard to estimate but it is not hard to be convinced that they are a small fraction of £60 per year. However, the tubular tyres of racing bikes are highly vulnerable, hard to repair and cost approximately £10 each. These thoroughbred structures have smaller safety margins in other ways too—a bump down, let alone up, a kerb can buckle a wheel noticeably and permanently. You will get knowledgeable discussion on these points.

Security is another important issue. Considering the fallibility of the security arrangements of many school bicycle sheds, and the essential ineffectiveness of locks against the dedicated thief, premiums for theft insurance remain incredibly low (about £1 per year), suggesting that most bikes do not go to school regularly! A discussion on improving security arrangements opens up another nice area, involving all sorts of planning skills.

All these considerations can be extended to a discussion of mopeds, motor bike or car ownership—already a lively subject for discussion amongst 16- and 17-year-old boys. The following are notes on one such session with a group of four boys; they had part-time jobs which gave them much wider financial horizons.

First they identified some of the main factors in buying a motorbike. 'It depends on what job you've got and what you earn.'

'And on your other commitments.'

'You've got to decide your priorities.'

'Long-term and short-term ones may be different.'

'You need other equipment to use the motorbike.'

'You could get a better job.'

'Yes, you could work extra hours to buy more equipment.'

'What's the satisfaction?'

'It all depends on the amount you use things. Sometimes you can hire them.'

'What's the cost going to be?'

In discussing their commitments and priorities, personal transport, saving for holidays and scheduling their school work provided the main themes, together with some commitment to saving for longer-term needs.

'Most part-time jobs pay about £8 a week.'

'For a holiday—well you've got to pay about £130.'

'School work comes first (!)—but it's very variable.'

'What about weekends? Oh, they're for me.'

'A motorbike—I'll give up something else while I'm buying it.'

'Or not go on holiday this year.'

'It gives you transport to school for the cost of an album.'

'We need some money in the bank.'

It became clear that the main satisfactions sought in owning a motorbike were freedom of movement and the possession of a status symbol—the latter depending very much on the quality of the bike. (There was no discussion of safety factors in this most dangerous of teenage pursuits.) One of the group regarded the financial problems of owning a motorbike as insuperable; the other three expressed firm commitment. One of them was also short of money but already had a moped and hopes of 'trading up' to a better bike. Another was explicitly saving up for this purpose while the third said he definitely wanted one. They looked at costs and produced the table in Figure 2.4. The insurance turned out to be even more than they feared. They then took a similar look at means of raising money—borrowing from parents, saving, a bank

Engine capacity	250 cc	125 cc	moped
Second-hand price	<£350	<£250	<£250
Comprehensive insurance	too expensive	£100	£80

FIGURE 2.4

loan and hire purchase were all considered but the last two were not seen as hopeful sources for the under-18s. They looked at the problem of raising £300 to cover the initial cost of the bike and a year's insurance. This amounted to two holidays. Saving £5 a week, or £20 a month, in a building society at $7\frac{1}{2}$ per cent you would get, they said,

$$20 + 20(1+x) + 20(1+x)^2 + \cdots$$

where $12x$ is the annual interest rate, so saving would take rather more than one year. If you could get a hire purchase loan you would spread it over two or three years, but pay more in interest.

Other running costs were mentioned, but not discussed very seriously—they were obviously regarded as small compared to the initial outlay and insurance premiums.

Holidays had been a recurring theme in the discussion, which now looked at them in more detail. Prices of chalets or hotels seemed too high and attention centred on youth hostelling or camping.

'You could buy a tent.'

'It's hardly worth it for two or three times a year.'

'I've used one every other week.'

'Could you afford to use it that often?'

For holidays, the possibility of hiring a tent was discussed and a list of other needs compiled. This brought a recognition that some things would have to be bought for use with a motorbike. The two lists are shown in Figure 2.5.

Tent	Motorbike
flysheet	helmet
sleeping bag	gloves
stove	jacket
dixies	waterproofs
torch	servicing
	petrol
	top-box

FIGURE 2.5

By the end of the session the group had organized which topics needed to be investigated further, what extra information was needed and how it might be obtained.

Borrowing money

Let's look at the problem of borrowing money. We won't go through this in very great detail but just try to indicate to you the steps through which the modelling process is likely to go. Let's choose a situation somewhat more precisely. Assume we are borrowing perhaps £100 to purchase some substantial object—it might be a guitar or a very cheap car or whatever. There are various ways of borrowing money and the first step, the first mathematical model, would probably be an enumeration of the possibilities. You could decide whether to borrow from a hire purchase company, or from a bank, or from a relation or friend, and there are other possibilities too. For each of these there is a variety of choices about how to repay the loan. The explicit enumeration of possibilities is a most important skill. Mathematics offers a general framework, a notation, which can often help to make things clearer—it is called graph theory. Figure 2.6 is a graph for the money-borrowing problem. It is a very simple form of graph which contains no closed loops, and is called a tree diagram for fairly obvious reasons. One might want to regard this step alone as a single pass around our modelling process. Its validation would consist in checking whether we have omitted any other important ways of borrowing money, or of repaying it.

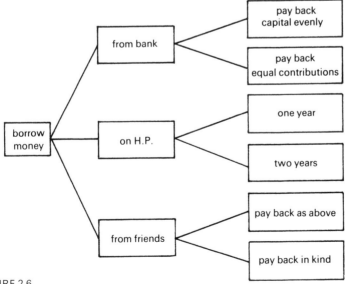

FIGURE 2.6

The next stage would probably be to discover how much it costs to borrow £100 by each of these routes. Obviously a lot of mathematics comes in here. It involves the calculation of interest and other things with which many will be familiar. Let us just notice that banks and hire purchase companies and friends each charge the interest on a different basis. A bank usually charges you interest only on the outstanding debt while a hire purchase company normally charges interest on the whole sum borrowed for the whole period. As is well known, this leads to quite different costs of borrowing. What then would be the specific mathematical questions you would be interested in asking? Probably, what is the total sum I would have to pay back, but not necessarily. If for instance you are living on a tight weekly budget it may indeed be the maximum payment in any one week that is the crucial determinant of how much money you can afford to borrow. The real-world situation determines the question that you wish to ask of the model and the criterion for what is a satisfactory answer. This seems a fairly cut and dried problem and having discovered what each is going to cost, you can then make a choice as to which method is preferred.

In fact there are several further stages which one would want to go through in this situation. First, inflation is a very important factor in our present times. The question as to whether one borrows money is greatly affected by the actual inflation rate. Traditionally the borrower repays the lender more in real terms than he borrows, the difference, the interest, being a fee for the use of the money in the period for which it was borrowed. However, in the last few years rates of interest have almost always been lower than the current rate of inflation so that the positions have been reversed, the lender has in fact been paying the borrower. In other terms, it has paid anybody in the last few years to borrow as much money as possible in the knowledge that he would be repaying less than he has borrowed. From an economist's point of view money could be used to buy assets which could be sold again at the end of the period for a higher sum than has been paid out in all, giving a net profit to the borrower and a loss to the saver. Of course, those who have tried selling second-hand cars or washing machines will know at once that this analysis is inadequate in this case; in fact they seem to depreciate at about the same rate as the rate of inflation. Those in a position to buy diamonds, Stradivarius violins or Rembrandts on borrowed money may have been in a position to make money out of it.

A further question concerns the availability of money from different sources. It is no good choosing the cheapest method if it is not one accessible to you; this factor, rightly or wrongly, is probably the main reason for the extensive use of hire purchase as a sort of borrowing.

All these, and probably some other factors, are an essential part of any realistic discussion of borrowing. Some may claim that these considerations are not part of the job of the mathematics teacher but I believe that he and his pupils will derive benefit from taking them on board.

What is money worth to you?

A teenager has 100p pocket money each week—his parents provide all the essentials including clothes and school dinner money, which we shall naively assume is spent on dinners, so that he is free to spend the pocket money as he chooses. He might buy any of:

7 Mars bars—good to eat in large but not excessive quantities
1 cinema ticket—excitement and pleasure
1 good ball-point pen—pleasure of efficiency and possession
3 Yo-yos—pleasure of a skill which is either unusual or fashionable
20 cigarettes—social pleasure in joining group, adopting adult behaviour, rejecting adult advice (leading to physical need)
2 pints of (illicit) beer—as above plus access to pub social environment plus disinhibiting effect
1 bottle of aftershave—improved self-image and social attraction
1 pop record—musical and social pleasure.

He could, alternatively, buy many other things, for example:

1 tin of floor polish
0.6 gallons of petrol
15 razor blades
£20 000 insurance against his death that week

but he would be much less likely to do so. Why, in each case?

In fact he would probably buy some of several of these commodities, the mixture being a matter of personal preference varying with the individual and the week. Clearly there are difficulties in comparing the pleasure of eating a Mars bar or perfecting a yo-yo trick, but for any two possible combinations of purchases to any value, he would be prepared to make a choice, gradually revealing his *preference function*—a function of the amounts of each commodity. Where is the pay-off in using mathematics here? First, we bring what is normally an instinctive and largely thoughtless process of individual buying decisions under scrutiny. Once questions have been raised, even though they may be thought ridiculous at first, they tend to stick in the mind. Many actions are motivated by a wish to fit into the pattern of the social group or a fear of seeming different—when these are recognized the individual will often assert his individuality, or at least economize. You can make a drink last all evening if social access is really what you want.

Secondly, this kind of analysis points out basic facts of economic behaviour that are essentially mathematical and rather sophisticated. That money is a single variable with a specific conversion rate, the price, into most other things (not happiness, of course). The preference function reveals your view of where prices are 'wrong'—for a child, the first few Mars bars are cheap—a much better buy than life (death) insurance, which to him is valueless. The 95th Mars bar that week, however, does not hold such attraction so that the 'proper price' depends on the quantity.

We'll come back to look at a simple two-variable version of this problem when we discuss optimization in Section 4.2. For the moment we merely suggest that a discussion of these problems, perhaps with the preparation of individual pocket money budgets by each child on the basis of different total sums would be a worthwhile exercise, giving, incidentally, practice in arithmetic and the organization of information.

Home budgeting—how much control?

The spending of pocket money is regarded by parents as good training in using money, and as a way to keep children quiet. The independent person has a more serious game to play in his spending decisions in that he could end up cold or hungry if he got them badly wrong. Home budgeting is a much more elaborate preference exercise with many more variables. The solution worked out by the housewife walking around the supermarket with a trolley, picking things off the shelves, is certainly not based on a thorough understanding of her complete preference function. How do we operate to solve that problem? Perhaps she works largely on the basis of last week's solution, with variations to meet some need or wish within the same money.

We have said enough about planning spending to make it clear that you could spend a lifetime thinking about this week's shopping, which doesn't seem very worthwhile. (It is a characteristic of applied mathematicians that they often run riot, far beyond the point of usefulness; this does no harm if you don't take it too seriously, and may reveal new possibilities.) The next real question is how much control of spending we *want*. We shall look at two examples—comparative shopping and, again, household budgeting.

Do you take your calculator to the supermarket?
The eager comparative shopper can check every buying decision. Even when he knows what commodities he wants, and in what quantity, the brand and size may be varied. Dividing cost by quantity for every size of your favourite brand of instant coffee, you find which is the cheapest—it isn't always the largest by any means. (More subtly, buying a less favourite brand may reduce consumption, which in these days of general overeating may not be a bad thing.) However, if you do this for everything you'll take all day. Ask the class how you should decide.

Suggestions will include: how much do you save, how much time does it take? Perhaps the question of how often you should check will be raised—it may be worthwhile once a month, or even once a year. Get pupils to work out a typical 'shopping basket' and guess how much could be saved by careful comparisons. Then, if possible, go in groups (with calculators) and work it out on the spot in the supermarket. What is mum's (or dad's) time worth? This opens up a lot of possibilities:

- nothing
- what she could earn ('but she wouldn't?')
- what else she could do to save money in the time (go to market, make something, grow vegetables, decorate house)
- relaxation.

Find maximum, minimum and 'reasonable' values, then work out how often this would justify checking prices, if the saving were multiplied by the number of weeks between checks. Do prices stay constant that long?

As usual, after a thorough exploration, find some conclusions and write them down. Are they obvious? Were they obvious before? If not, there was a pay-off. Now let them try to persuade mum.

Household budgeting needs similar justification—if you check every spending decision you'll never do anything else. But if you never think about it, you will spend money on things you want less than some other items you will believe you can't afford—unless you're very rich. How do *you* work? Is it a good enough method? Most people seem to have a small maximum sum up to which they buy what they want without thinking—provided, at least, it is not obviously going to recur too often. Clearly the larger the expenditure the more carefully it needs to be considered. Get the class, working in small groups, to list the things that a family might spend money on. Then for each item or group of items, work out how the decision whether to buy, and how much to spend, should be made. For each group get them to justify one such recipe to the class as a whole.

The ideas of optimization, of making the best of a situation, are central to this and many other bits of practical applied mathematics (see Section 4.2).

Ellen's horse sense

Ellen is 13 and keen on horses. She lives in the country and has learned to ride. She has even had some success in jumping competitions on other peoples' horses that she has been allowed to ride. She wants a horse of her own for many good and obvious reasons but she has been told by her parents that keeping a horse is far too expensive. With the new idea of sharing the horse with a friend, she returns to the attack. (Figure 2.7)

Pony Problem.

Food amount and cost.
HAY - For a 14·2h horse we will a ton of hay a year
There are 45-50 bales of hay in a ton at 60p per bale
1 ton = £27·00 . 12 months a year = £2·25 per month

PONY NUTS- 4lbs of nuts a day for 7months.
½ cwt of nuts = £3·00 which will last for 14 days
£6 per month for 7 months = £42 per year
12 month a year = £3·50 per month

OATS - 4lbs of oats a day for 7 months
½ cwt of oats = £3·00 which lasts for 14 days
£6 per month for 7 month = £42 a year
12 months a year = £3·50 per month

FIELD = £2·00 a week
£8·00 a month

Insurance £2·50 a month

VET = £2·50 a month

SHOES = £3·50 for removes
£5·50 for new shoes
need removes 6 times a year
need new 3 times a year
12 months a year = £2·90

total cost =

HAY	–	£2·25	approx	per month
NUTS	–	£3·50	approx	per month
OATS	–	£3·50	approx	per month
FIELD	–	£8·00	approx	per month
INSURANCE	–	£2·50	approx	per month
VET	–	£2·50	approx	per month
SHOES	–	£3·50	approx	per month.

£25·75 total a month between 2 familys
£13·00p per month each
£3·25 per week each

FIGURE 2.7

Not too unreasonable compared with holidays, say. It's almost within pocket money range if enough brothers and sisters join in—but how many can a horse 'carry'? So much for money; now for time and people. (Figure 2.8)

Pony Problem.

Possible Field Owners. Wilcox's   Harlands
                       Owens   Macintoshes
                       Eggins   Skinners
                       Doidges   Madgewicks
                              Others.

Time-table for any winter school-day between December + April

Get Up	Dress	Go to Horse	Feed him	Go home	Breakfast	Get ready for school
6·45	7·00	7·15	7·30	7·45	8·00	8·30

School Hours	Horse Hours	Tea	Music /Homework
8·30 - 4·30	4·30 - 5·30	5·30-6·30	6·30 - 9·00

we would get 9½ hours sleep.

FIGURE 2.8

Who would bet against their winning?

What do you want from the 'pools'?

The football pools, and other gambling situations, are each an illustration of combinatorial mathematics and, particularly, of probability theory. Here we want to raise in a rather explicit form the more general question of what one is buying when one sends in the coupon each week.

A football pool syndicate is a group of people, N of them say, who share the 'investment' so that they can predict more possible sets of results and so increase the chance that they will win a 'big prize' of say £100 000. The chance of this, which will in any case be very small because the pools make money and pay taxes, will be proportional to the number N of people who contribute their stake of perhaps 50p a week to the syndicate and of course divide the prize. Let us discuss the situation for different sizes of syndicate so that each of us can decide what size of N he would prefer.

Many different approaches are possible and, provided the class has some simple notions of probability, an open discussion should bring out many of them. The more microscopic approaches, involving the chance of predicting eight draws out of fifty matches, provide valuable exercise in the multiplication and addition of probabilities; they are likely to be very unrealistic because the results for all matches are not random. Indeed, in most weeks there is no large prize given out because too many people make the optimum prediction. (The premium bond is the better arena for this sort of exercise but they do not seem to arouse much interest.) A macroscopic approach is more likely to be successful. It should be clear that we are only interested in rough orders of magnitude.

Since the 'pools' are profit-making and tax-paying it is clear that the chance P of winning a £100 000 prize from a 50p stake must be about

$$P_1 < \frac{0.50}{100\,000} \approx 2 \times 10^{-6} = \frac{1}{500\,000}.$$

With N people in the syndicate the chance will be increased and the prize V divided proportionately so that

$$P_N = NP_1 \quad \text{and} \quad V_N = £\frac{100\,000}{N}.$$

So as N is increased, your chance of winning increases but the prize itself gets smaller. In the table below various values are given for N, P_N, the number of years before you are 'likely' to win, the size of the prize, and a purely subjective view of that size of syndicate.

N	P_N	'Years to wait'	V	
1	2×10^{-6}	$\approx 10\,000$	£100 000	Could really change your life style
10	2×10^{-5}	≈ 1000	£10 000	Could get some long-term comforts or make a big splash
100	2×10^{-4}	≈ 100	£1000	Christmas really comes but once a lifetime

It would appear that very large syndicates with small prizes do not make much sense because the fluctuations are averaging out, and in the long-run the punter is sure to lose money. (The large number of people who continue to gamble frequently for small prizes makes it clear that this is not a general view—presumably they are deriving some satisfaction from the actual winning experience itself. How do your class feel about this?) On the other hand the *hope*, however small the chance may be, of winning such a large sum of money that one's whole style of life would be altered is something that requires N to be fairly small. When the average industrial wage is in the region of £4000 to £5000 a year a sum in excess of £20 000 seems necessary, so that the syndicate must not contain more than a few people. In between, one is purchasing the larger hope of a smaller windfall of a few hundred to a few thousand pounds, which may also seem a reasonable objective.

Time is short too

There are many situations where time allocation problems arise. The response of the children in 1E, reported on page 19, shows that many children are worried about coping with homework adequately without spoiling their social life.

2.2   Living together

Domestic life provides a rich variety of situations, some complex, some with an element of conflict, that repay analysis using some mathematics. This chapter discusses a few of them—your classes will give you many more to intrigue you and to challenge your problem-solving skills—if you have doubts about these, give the problem back to the class. Common sense will get you all some distance and you will find that mathematical skills have usually been used, too. Sensitive issues will occasionally arise and care must be taken to handle these sympathetically.

Decorating

Among the hallowed examples from the past, wallpaper calculations held an honourable place—how many rolls (11 yards by 21 inches) would be needed to cover the walls of a room 12'6" by 10'9"? The model used was not very good, but it provided good practice in areas and *some* feeling of reality. However, there is much more to home decorating than this and it is an Action subject for many teenagers, at least as far as the decoration of their own room is concerned.

A discussion may produce a list of ideas like Figure 2.9.

one room	walls	Cost
paint-emulsion	ceiling	areas a, b, c...
- gloss	door	quantities of paint L litres
wallpaper	windows	cost per litre £p/L
coverage	paintwork	no. rolls of wallpaper N
posters		cost per roll £w
brushes, sandpaper etc.		dimensions of room.
time ?		height of picture rail
sources of money		

FIGURE 2.9

The suggestion to work out the cost of decorating the room in various ways is likely to follow. Each pupil can do it for his own room. The basic cost of two coats of emulsion paint on walls and ceiling with one of undercoat and one of gloss on the paintwork involves working out areas and the quantity of paint needed to cover a given area (there is an estimate of this on most paint tins—approximately $3\,\mathrm{m^2}$/litre depending on the paint and the surface) and the cost per litre. Incidental exercise in proportional reasoning is provided.

The wallpapering alternative is *not* an area calculation but involves length. (Most people outside mathematics textbooks do not want joins in the middle of vertical strips.) We must first find out how many vertical strips we can get out of one 10 m roll. Each vertical strip must be longer than the height to be covered, from skirting board to picture rail or ceiling, and must be a whole number of pattern repetitions. If, for example, this length is 2.2 m then we get four strips out of each roll; the 1.2 m bits left over will be useful over and under windows, etc. The second length calculation is to work out the perimeter of the room by adding together the lengths of the walls; dividing by the width of the roll gives roughly the number of strips needed. However, we need to see how many rolls can be saved by using the odd lengths off the ends of the roll in the odd bits of wall above or below windows or doors. The cost of wallpaper varies enormously from about £1 a roll upwards. Some investigation of local decorating shops will give pupils an idea of what they get for their money.

It will emerge that the wallpaper only replaces the painting of the walls, so that it is convenient to calculate:

·   – total cost of painting room
     – extra cost for wallpapering at £1 per roll
     – extra cost for wallpapering at £2 per roll

and to compare these last figures with the cost of posters or other decorative objects that might be used on plain walls.

Sources of money can also be discussed. Parents may well pay for materials if labour is provided, and trusted, so that within reasonable bounds the cost of material may not be crucial. On the other hand, a lump sum might be negotiated, allowing the student to profit from his ingenuity.

The relative life-time of different forms of decoration, under various patterns of wear, may also be estimated.

The cost of equipment for decorating can be quite significant. Although many cheap brushes are obtainable, they have their disadvantages. Professional equipment for wallpapering (table, wallpaper brush and seam roller) is even more expensive. These may sometimes be hired, or borrowed from the shop selling the wallpaper, though the cost of this free service may be reflected in rather higher wallpaper prices. The comparison between the costs of renting and buying is a fairly general problem, applying also to television sets, camping equipment, or housing.

Helping in the house

Sharing out the jobs is another aspect of living together that gives rise to friction. There may be a few homes left where mother expects to do everything except mow the lawn, or where the roles are so well-defined and specialized that there is no question as to who does what and when. Otherwise there are division and scheduling problems to be sorted out; these may be solved informally and amicably, but in other cases there is a growing feeling of resentment by one (or all) of those involved. Even if it is only laying tables, clearing and washing-up after meals there may be a pay-off in looking at the situation.

A discussion of the whole problem will reveal a range of patterns and problems. From these will emerge factors like:
- a list of jobs
- who shares which
- preferences and other commitments
- payments for some jobs.

Drafting schedules for the main jobs in their own household should be straightforward for most children. It will probably lead to a recognition of the over-rigidity of such a system and escape routes can be discussed such as swapping jobs, allowing so many to be missed per week, etc. Are paid jobs compulsory? How are the parents included? What sanctions are there for regular shirking? Do you want a system like this in your house? Would mum like it?

The mathematical demands are simple—the organization of information in tables, arithmetic, and a trial and error approach to fitting in cycles of children and jobs into a week. These could lead to more mathematical questions about the compatability and periodicity of such cycles, though the pay-off from answering these in a more general way is doubtful. Figure 2.10 illustrates such an approach. Pupils will fill it in with an interesting variety of patterns; it is worthwhile for them to compare their solutions.

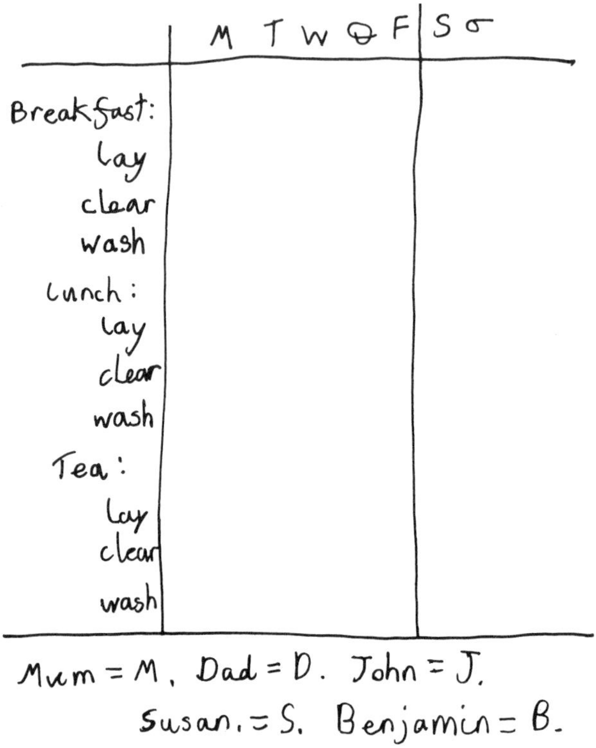

FIGURE 2.10

Clothes and fashion

Clothes provide another vein of interest and are of financial importance to teenagers. The competing demands of fashionable variety and shortage of money provide an optimization problem which might gain something from a thoughtful analysis. We have to admit, however, that we have no very startling insights to offer.

First there is the comparative shopping problem. Jeans, the key standard fashion garment, vary in price by at least a factor of 3 even for this age group, from those on the stalls at the local market (£6 in 1978) to the famous American names, Levi's or Wrangler's, at about £18.

Fashion is the key determinant and a discussion of all the factors (a *Which?* for jeans) should include wear, fading and other signs of decay in conventional clothes which may (or may not) be assets for jeans provided they remain useable. There is little doubt that the more expensive jeans are made of heavier denim, which may affect matters either way.

Secondly, the central place of jeans as a uniform garment (most of the arguments in favour of school uniform would be met if jeans were compulsory) means that they play a different role to shirts, sweaters and other top garments—these are changed frequently so many more are needed and the average unit cost needs to be kept down. Indeed the fashion allows a remarkable variety including grandfather's shirts—without collars—and boyfriends' rugby shirts; the current demands by girls for these suggests a sensible financially practical side to fashion trends.

The third factor is the source—are the garments bought by teenagers, by parents as gifts, or what? The simplest device—a list of clothes bought (and sold?) over a year with prices would certainly give interest and surprise—a sample survey of how often each garment is worn would yield more information on which to base further discussion. These are sensitive areas, particularly for less privileged children, who are not always those from the hardest financial backgrounds. They should not be pushed hard in class.

Other aspects of fashion may be of strong interest to some. Those who make their own clothes face a lot of interesting problems such as:
- How much do you save on various types of garments?
- So how much is your time worth?
- Can you squeeze a pattern out of less material than it asks for?
- Is a sewing machine a good investment?
- Can you adapt interesting second-hand garments?

These are things where a good deal of technical knowledge is useful and are likely to be discussed in a class where both the teacher and some of the pupils are actively involved in dressmaking.

Music, music, music...

Music is a central feature of the lives of many of us, and most teenagers. Are there any questions or problems that arise? A discussion may well yield something like Figure 2.11.

We shall choose to discuss a few of these questions. We leave out the buying of equipment because the comparative shopping for consumer goods is fairly straightforward, though deciding the quality of sound reproduction you want, and whether a given 'set-up' gives it, are interesting questions.

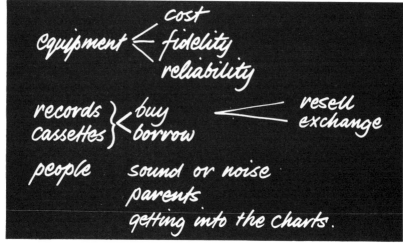

FIGURE 2.11

In choosing between records and tape we may consider:
- relative cost
- 'singles' or 'albums'
- storage
- convenience, use and portability
- exchange or resale
- other uses.

Tape can also be used for personal recording as well as for copying other people's records or radio programmes, though this requires a record player or radio as well and may infringe the copyright laws. Each of these aspects can be 'costed' and such a discussion may provide real help with decisions.

Sound insulation is often a problem. Your music and enjoyment may well be other people's noise and discomfort. How good the insulation must be depends on:
- how loud the noise is
- how much background sound there is in your area
- whether you are sitting reading, watching TV or running a disco
- how annoying you find the sound that reaches you.

In discussing sound insulation we are concerned with the physics of sound transmission and the physiology of the ear. A thorough treatment of these things is beyond general knowledge but there are a few guiding principles.
- Walls are good sound insulators.
- Ceilings and floors are moderate sound insulators, with more sound going down than up because of loudspeakers resting on floors or furniture.

  – Doors are bad sound insulators, particularly open doors!
  – Two layers of sound insulation are much better than one.
  – A sound can be 'masked' by another louder one.
So for making loud noises you should choose:
  – not to be in an adjacent room, certainly not overhead
  – to keep the doors closed, both of the room where the sound is loud and in the 'quiet room'
  – to have some acceptable background noise in the 'quiet room'.
Look at your house and consider how the rooms might be used to allow loud music with minimum disturbance, and get each pupil to do the same thing. To be definite get them to consider how best they might organize:
  – a disco in their room while parents watch TV
  – quiet working in their room while parents watch TV
  – two children, one working, while the other has friends in and music on.
You may like to introduce some 'solid' teaching of information sometimes, to provide the background to enable the pupils to build a better model of the situation. Generally speaking, such teaching needs to be planned and prepared ahead of time. In the present case we might choose to talk about how loudness is measured, the ear's logarithmic sensitivity to sound intensity and the scale on which it is measured.

  If a sound is suddenly increased in magnitude, the listener receives an impression of increased loudness which is roughly proportional to the logarithm of the ratio of the two acoustical powers, i.e. δ (loudness) $\propto \log_{10}(P_2/P_1)$. The *bel* is the measure of change in loudness and the number of bels is defined as $\log_{10}(P_2/P_1)$. But this is inconveniently large and the commonly quoted unit is the *decibel*, ten to the bel.

  The following information might be appropriate.

Some sound levels	
Threshold of hearing	0 dB
Leaves rustling in the wind	10 dB
Alarm clock ringing at 1 a.m.	80 dB
Maximum 'safe' level for a working day	90 dB
Disco or pop group	110 dB
Concorde taking off	120 dB
Threshold of pain	120 dB
Dangerous to hearing	135 dB

  Those of us who have read the sales literature for hi-fi amplifiers will have encountered decibels. Suppose we have an amplifier with an output of 5 W and we are tempted to go for something appreciably louder, and are attracted by an advertisement for an 8 W amplifier. Will we notice the difference? The change in loudness would be $10\log_{10}(8/5) \simeq +2\,\text{dB}$. Now a change of one decibel is hardly perceptible to the ear, and 2 dB only just so. A 10 W amplifier would give a perceptible rise of about 3 dB and 15 W a rise of about 4.8 dB. But note that this is not a measure of intensity of sound, but simply a useful factor for comparison.

If the threshold of hearing is taken as 0 dB, then the threshold of pain is about +120 dB. But on the common electrical multi-meters there are arbitrary reference levels for 0 dB: many American instruments specify, e.g. 0 dB = 6 mW or 1.73 V into 500 Ω, and a well-known British instrument maker specifies 50 mW = 0 dB. The British instrument will then constantly read just over 9 decibels below the American reading.

'Hits' in the charts may be worth investigating, though mainly to satisfy curiosity. A study of the progress of a particular record from week to week can reveal quite different patterns of behaviour—in some cases there is a meteoric rise and fall while in others the record has a long period of comfortable popularity. Children have ideas as to how this depends on the fame of the song or the group concerned, and how it affects the total sales of the record, though this requires collecting further data from the musical press. Although there is little likelihood that these questions will form Action problems for the children in the class, their ability to identify with the pop music world helps to make them highly Believable.

A domestic postscript

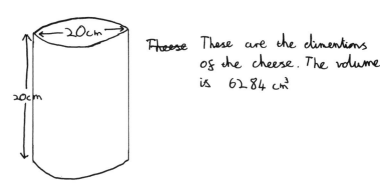

The Problem About The Cheese

These are the dimentions of the cheese. The volume is 6284 cm³

This cheese can be cut in two ways:

sliced   or wedged

  and it is cut into eight pieces at the end
The problem is to find which is the better way
to cut it so it reduces the amount of drying
up (by reducing surface area). The surface
area at the beginning is 1884 cm².

We shall look at the sliced first. When one
slice is cut off, 471 cm² of surface area is lost
and 314 cm² of surface area is gained. This
leaves 1727 cm² of surface area. Of course, the
cheese is cut more than once and so the results
are shown on the next page.

  Now, we will look at the wedged. When
one wedge is cut off, 235·5 cm² of surface area
is lost and 400 cm² of surface area gained. This
leaves 2048·5 cm² of surface area, which is more
than when we started. But as the table below shows
it does come down again.

AMOUNT OF CHEESE	$\frac{8}{8}$	$\frac{7}{8}$	$\frac{6}{8}$	$\frac{5}{8}$
WEDGE	1884 cm²	2048·5 cm²	1813 cm²	1656 cm²
SLICE	1884 cm²	1727 cm²	1566 cm²	1413 cm²

AMOUNT OF CHEESE	$\frac{4}{8}$	$\frac{3}{8}$	$\frac{2}{8}$	$\frac{1}{8}$
WEDGE	1342 cm²	1104.5 cm²	871 cm²	635.5 cm²
SLICE	1256 cm²	1099 cm²	942 cm²	785 cm²

These results show that the sliced has a smaller surface area than the wedged except for at the end where the wedged overtakes. So, really, I think it is better to slice the cheese, one slice at a time, rather than cut it into wedges.

FIGURE 2.12

2.3   Movement

Mechanics traditionally *was* applied mathematics, but it has suffered a steady decline in mathematics syllabuses. In some ways this has not been a bad thing, since mechanics was often taught in mathematics in a way which had little contact with the actual phenomena and blurred the distinction between the purely mathematical statements and operations, and the assumptions of the model—the choice of variables, Newton's physical laws and the prescription for the forces in each case. Also, mechanical phenomena are usually discussed quite carefully in CSE and O-Level science courses as well as in physics courses, and there is a case for saying that the models are sufficiently difficult in concept to merit only specialist treatment.

However, movement does fill our lives: we, our friends and relations, other people and things, the sun, the wind are all it seems in motion. The two ingredients for understanding these phenomena are the geometry of spatial relationships and the dynamical laws of interaction between things, crystallized by Newton; neither is easy to grasp but both are best expressed with the help of mathematics. The range of interesting phenomena that demand an explanation is so wide that every child should get some feeling for the power that mathematics provides in understanding them.

Conversely, mathematics can benefit from the enthusiasm of many children for engineering—mechanical devices which perform a useful function do not receive much attention in the school curriculum but their design usually involves a variety of mathematical techniques from geometry as well as from arithmetic and algebra. A good deal of pioneer work has already been done on the essential task of bringing realism, including practical experiments, into this work. Some of it is discussed in two reports prepared for the Mathematical Association, *The Teaching of Mechanics in Schools* (1965) and *Mathematics Laboratories in Schools* (1968). The Mathematics for the Majority Project also opened up a range of exciting ideas, see Figure 2.13—taken from *Machines, Mechanisms and Mathematics* (Chatto and Windus, 1972).

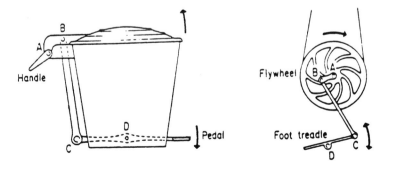

FIGURE 2.13

It seems to us sensible to base the study of mechanics on the phenomena of everyday life and particularly on those areas of keen interest to many teenagers—traffic and sport.

Leaving traffic lights

The movement of road traffic provides a rich variety of problems, a good proportion of which lie within the experience of this age group, as observers if not as participants. A nice example is the way a queue of cars, stopped at a red traffic light, moves when the lights change to green. Each car starts a second or two after the one ahead has begun to move away. If all the vehicles are of the same type this works very smoothly since they accelerate up to cruising speed in a similar way, the car that is ahead is always moving faster than the one behind and so the separation between the two of them increases. But it does lead to a rather slow take-off from the traffic lights, with a wave of 'starting' moving back down the queue at not much more than a brisk walking pace. Could we do better?

Let's first try to describe mathematically what happens normally. We use algebra because of its greater clarity, but will choose specific sample values for each parameter so we have a clear quantitative idea of what is going on at each stage. The whole analysis could be done perfectly well with arithmetic only, but one would probably want to work out a number of cases; we shall work out one later for illustration.

Each car starts to move at a time $t \approx 2$ seconds after the car ahead moves. If the distance between the front bumpers of succeeding cars at rest is $l \approx 6$ metres, then the car number n at distance x from the lights starts moving $nt = (x/l)t$ seconds after the lights change to green. Let us assume each car accelerates quickly up to $v = 10\,\mathrm{m\,s}^{-1}$ (about 20 mph) and then goes at a steady speed, then the gap between cars will be increased by $vt \approx 20\,\mathrm{m}$.

To understand this result, notice first that if all the cars start at the same instant and accelerate to speed v in the same way, their separations will remain constant. Each takes the same time and travels the same distance in getting up to speed. The interval t between two successive cars starting means that, once they have both reached speed v, the first has travelled t seconds longer at this speed giving the increase in separation. We shall illustrate this graphically in a moment.

How many cars will get through in the time $T = 40\,\mathrm{s}$, say, for which the lights stay at green? If a car starts a distance x back from the light, then the time from when the lights change to green until the car reaches the lights can be thought of as the sum of the waiting time, the time it would take to travel the distance to the lights at a steady speed v and an extra time t_a to allow for the time lost while the car accelerated up to v. So

$$T = nt + \frac{x}{v} + t_a \qquad (2.1)$$

which we can solve for $n = x/l$. It only remains to estimate t_a. It takes, say, 4 seconds of steady acceleration to reach $10\,\mathrm{m\,s}^{-1}$. The average velocity during this period is $5\,\mathrm{m\,s}^{-1}$; compared with the steady velocity v, therefore, an extra $t_a = 2\,\mathrm{s}$ are occupied over the distance covered in this phase. Solving equation (2.1) we find

$$n = \frac{T - t_a}{t + l/v} \approx \frac{40 - 2}{2 + 6/10} \approx 15.$$

So, interpreting, 15 cars get through. (Note: t_a is a small effect.)

Could we do better than this with another system of traffic organization? A number of possibilities have been explored theoretically but one idea is this. Suppose drivers were trained to park with about a 2 metre gap between their cars, corresponding roughly to our $l \approx 6\,\mathrm{m}$, and all to let in their clutches simultaneously when the traffic light ahead turned to green (the lights might need to be placed higher in the air so they could be clearly seen back down the queue). If we assume that the pattern of acceleration is the same as before then equation (2.1) is replaced by one with no waiting time and

$$T = \frac{x}{v} + t_a$$

$$n = \frac{T - t_a}{l/v} \approx \frac{40 - 2}{6/10} \approx 63 \text{ cars.}$$

So in this case over 60 cars (four times as many) get through.

The hazards of this pattern of driving are clear, but there is a massive increase in the number of cars crossing the light at the green phase. This suggests the possibility of modifying the chosen values of some of our variables to increase safety whilst still allowing a big increase on current throughputs—a larger separation between the waiting cars and a gentler acceleration are the two most obvious possibilities.

This situation shows how a model may be useful in guiding a compromise between competing objectives—in this case, safety and speed of travel. The choice of the criterion for an optimum solution lies outside the model. In order to illustrate explicitly how the same problem can be tackled at different technical levels, we shall now briefly repeat the analysis using *only* arithmetic with *roughly similar* assumptions.

Assume that the cars are parked every $6\,\mathrm{m}$, starting to move at intervals of $2\,\mathrm{s}$ and accelerating 'quickly' up to $10\,\mathrm{m\,s}^{-1}$. Figure 2.14 shows a speed–time graph of the second and third cars in the queue; it is

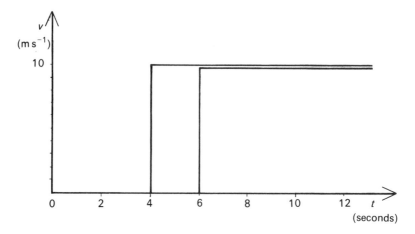

FIGURE 2.14

assumed that the lights turned green at $t = 0$, so that the first car reached the lights after just a 2 second delay in starting.

The second car:
- starts 2 seconds later, i.e. after 4 seconds
- takes $6/10 = 0.6$ seconds to reach the lights (this is obviously not good enough but we can come back and improve it).

The third car:
- starts after 6 seconds
- takes $12/10 = 1.2$ seconds to travel.

The fourth car:
- starts after 8 seconds
- takes $18/10 = 1.8$ seconds to travel.

We can now make a table.

Place of car in queue	Delay in starting (seconds)	Travel time (seconds)	Total time to reach light (seconds)
1	2	0	2.0
2	4	$6/10 = 0.6$	4.6
3	6	$12/10 = 1.2$	7.2
4	8	$18/10 = 1.8$	9.8
5	10	$24/10 = 2.4$	12.4
10	20	$54/10 = 5.4$	25.4
15	30	$84/10 = 8.4$	38.4
20	40	$114/10 = 11.4$	51.4

We see that the cars reach the lights 2.6 seconds apart and that 2 seconds of this is accounted for by the reaction time delay in starting so that this, rather than travel time, is what most seriously restricts the flow.

We have ignored the time to accelerate up to $10\,\mathrm{m\,s^{-1}}$. How can we take this into account and what difference does it make? We could plot the speed–time graph for a typical car in the queue in a more realistic fashion; Figure 2.15 is a very simple constant acceleration model, while Figure 2.16 tries to be a little more realistic (it is a good exercise to interpret what the features of this graph are supposed to represent).

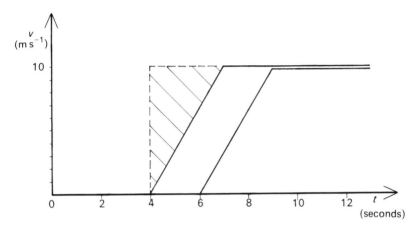

FIGURE 2.15

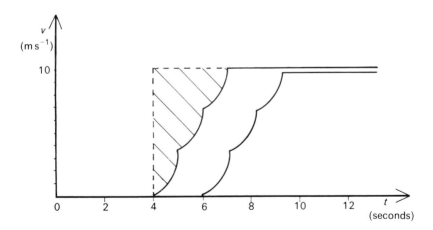

FIGURE 2.16

The broken line represents the simple model we have used so far. It is important to know that the areas under the curve on a speed–time graph represent distance travelled. Thus for any car reaching cruising speed before the lights, the simple instant acceleration model predicts that it has travelled farther than the better models by a distance equal to the shaded area; this is $\frac{1}{2} \times 10 \times 3 = 15$ m in the triangular case, Figure 2.15. So the car can be thought of as starting 15 metres further back and $15/10 = 1.5$ s must be added to *each travel time*—a small correction. The area is more difficult to compute in the case of Figure 2.16, but again the delay is the *same for each car*.

Obviously the situation is more complicated for the cars at the front which reach the lights while still accelerating, but they will always get through the lights anyway so only curiosity leads us to look at them further. In the constant acceleration case (Figure 2.15) the second car reaches the lights after a travel time such that the area under the curve is 6 metres. Algebra always helps such implicit problems—in this case the area after travel time t is

$$\frac{1}{2} \times t \times \frac{10}{3} \times t = 6$$

so the time $t = \sqrt{(36/10)} \approx 2$ s. If we wish to avoid algebra we could again tabulate distances travelled after various times and read off the times for various distances.

The method so far described is based on a mixture of memory and guesswork. In order to validate it, we should have to go out and observe what actually happens at traffic lights with a keener eye, verifying that our model is generally correct, measuring the typical values of the variables we have used, and noting the effects of variations in the standard situation. For example, the inclusion of a slow lorry or turning vehicles in the traffic stream will modify the analysis. A practical experiment to test the new method would arouse widespread interest, and not only for mathematicians.

For those who regard these ideas as completely crazy, we would only say that there exist (in the literature) suggestions which are apparently even more hazardous. One of these visualizes a second set of traffic lights about 100 metres back from the main lights on each of the entry roads from which traffic is released, so that it reaches the main intersection at full speed just as the lights turn green. This suggestion can be analysed in a similar way.

Traffic provides many other important situations which are essentially one-dimensional and reasonably accessible to some analysis. Observation of speeds, acceleration and braking of a car (What does the MoT test demand?) allow discussion of important driving tactics. The safe driving distance behind another car depends on how safe you want to be. If you assume you must be able to stop behind the car ahead when it brakes hard, and you react similarly one second later, you get roughly the 'one car length for each 10 miles per hour' rule. However, if you want to be able to stop if the car ahead has an accident and hits a bridge, a vastly greater separation (the stopping distance) is needed.

Overtaking has at least two essential elements—the time to accelerate and the time to pass. Which is normally the longer? The former can be 'absorbed' by accelerating in preparation for a possible opportunity to overtake—does this help? Only the passing time depends on the length of the vehicle being overtaken.

Traffic is thus a very rich source of problems in movement and mechanics. It has the advantage that simple one-dimensional kinematics is adequate for many interesting effects. It can be approached in the classroom either from the real situation or in a more conventional didactic way, using realistic illustrations. Appendix B gives a fully worked out teaching unit showing this alternative approach.

The other obvious area where movement enters importantly is sport. Its possibilities are just beginning to be explored but it seems to have fewer interesting-but-not-obvious results from very simple analyses.

Football

Football is above all a game of movement, with a number of different speeds. When a side recovers possession of the ball, there is a period of repositioning as each readjusts from defence to attack, or vice versa. Then a phase of build-up as the ball and players move rather slowly up, and across, the field. Finally, the attack with players moving at top speed. Over it all hovers the potential speed of the ball, an order of magnitude faster than the men, but subject to error in direction.

The alternative fast break into attack is used, sometimes to crucial effect, when there seems to be a chance to catch a 'stretched' opposing team before they can consolidate their defence. The slow build-up is used more often in higher standards of football where the chance of slipping through the defence near their goal is less and controlled possession has a higher priority than 'getting the ball up the field'.

A significant observation about football is that attacks always fail, more or less, (scores are low, like 2-1). Why? In top-class basketball attacks generally succeed (scores like 110-106 are common). Can efficient defence always defeat attack? Why? What are the optimum strategies for each to pursue? Is tactical skill important or does everyone know what to do and is it only technical ability to do it that is the limitation? What are the main sources of breakdown—error in direction or velocity of passes, bad communication between the players? Should a man dribbling the ball expect to beat a defender? Is ball control or acceleration the limitation? These are ambitious but seductive targets to study in a mathematics lesson. We shall look at a few of the simpler ones.

It would be helpful if, before embarking on an analysis of football or any other athletic activity, you can discuss it with the physical education staff of the school. Explain you wish to make mathematics more useful to the children and seek their views on the issues discussed in this section. Then, if later one of the pupils raises the matter in the gym or on the sports field, he may get a positive reinforcing response.

Speed of ball and man

Among the basic variables in the game are the speed at which the ball and the players move in various phases of the game. The speed of the free kick, the slow build-up, the burst through the defence, the through ball—what does each of them involve? In passing a ball cross-field to a running player how far ahead of him should it be aimed? (This is the same vector problem as on page 5. Get the class to guess—they may say 5 yards perhaps.)

How can we estimate the speed of a ball?

1 *In a long pass*—Look at football on television. Have several pieces of paper marked out with a scale drawing of a football field. When you see a pass, mark on the paper roughly where it begins and ends and time roughly how long it takes. You can get such timing either by calling 'now', at the beginning and the end of the pass, to a helper with a second hand on his watch, or by counting seconds. Someone in the class may even have a digital stop-watch. The calculation

$$\text{speed} = \frac{\text{distance measured on 'map'}}{\text{time}}$$

for a selection of long passes will give an estimate of speed. Corner kicks and 'crosses' are similar—or are they?

2 *Goal kicks* can be done the same way—how often do they 'give the ball away', and why?

3 *Penalties, free kicks, and shots at goal* clearly move faster. How could we measure their speed? Any direct time estimate would be very rough but might be worth doing all the same. We could do much better either with a film, which we know has 16 frames a second (or 24 with sound), or via the 'action replay' on television. We can find how much this is slowed down from the player's run.

The speed of the player can be estimated similarly—it will also vary a lot—and it can be checked from our knowledge of how fast athletes can run (about $10 \, \text{m s}^{-1}$ over short distances).

Goalkeeping

The goalkeeper in modern football has a very varied role. He is, of course, unique in being able to use his hands and we discuss the direct consequences of this in terms of reach and general coverage, but this is but one aspect of the many spatial problems that the goalkeeper must solve. Let us start by discussing another—'narrowing the angle' or, as we would rather call it, 'covering the goal'.

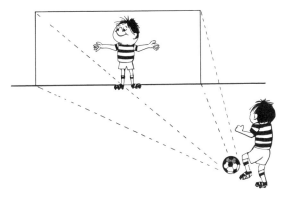

FIGURE 2.17

When an approaching forward runs in on the goal, perhaps as in Figure 2.17 the goalkeeper must decide how to position himself. If he waits on his goal line the forward has a large area to shoot at which, because the shot moves much faster than the goalkeeper, means he is likely to score. The goalkeeper must therefore try to reduce the area of goal open to the forward—he must try to block the goal so that there is no line that the ball can follow from the boot into the goal. He can do this by coming towards the forward—indeed if he can reach the ball he can smother the shot completely—but this takes time and leaves the goal unguarded from other directions.

FIGURE 2.18

Often the goalkeeper will not be able to get far enough completely to smother the ball before the forward shoots. How can he cover the goal most effectively? Normally he is not a very good mask for the goal because a man is, more or less, a tall narrow object, while the goal is a short wide one (see Figure 2.18) so that the shadow of the goalkeeper will not mask the goal very effectively. He can do this much better if he can rotate his body so that it is lying horizontal at an appropriate distance above the ground. (Figure 2.19)

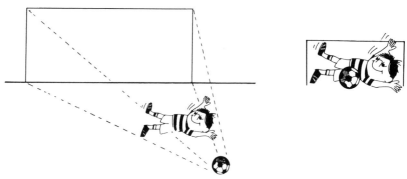

FIGURE 2.19

Good goalkeepers do just this, flinging themselves on their side at the moment when the forward shoots. It is called 'spreading'. Timing is obviously very important as this is not a position that can be sustained! This is but one aspect of the geometry of goalkeeping.

A similar analysis gives different conclusions in hockey, where the stick provides an extension to the goalkeeper's arm and so makes the effective shape much wider; this difference affects many aspects of the two games, including coverage by defenders.

A postscript on goalkeeping—forwards are trained mostly to shoot low and to head down because the goalkeeper can 'go up' more quickly than he can 'get down' (slip fielders in cricket crouch on the same basis). Ask the class why this is so and how it could be tested.

Errors and accuracy

Another crucial factor in skill at football is the accuracy with which the ball can be kicked. This again is something that can be estimated in the classroom and then checked experimentally. Some of the boys in the class may take 20 shots each with a stationary ball, and later with a moving ball, at a target marked on the wall and the distribution of their shots noted and plotted, preferably in terms of the error in the angle of the shot. The variation in the error from shot to shot for each player, and the variation in the mean errors between players, are worth discussing.

Defensive coverage

A long pass also gives the defender more time. The general principle of marking is based on this—the farther away the ball, the farther from 'your man' you can afford to be. It allows the defender partly to move over to a territorial 'zone marking' when the ball is far away, closing in on his man as it approaches.

How big a zone can a defender cover? This will depend on how far away the attacker with the ball is, how fast the defender can move and react, and how fast the ball travels. With chosen values for these parameters, the geometry is quite straightforward. You can lead on to a deeper discussion of defensive tactics. The picture will break down when the two players are face to face—a more difficult situation to analyse.

It is the essence of good defence to make your opponent do what *you* decide; to control his options, leaving a wider gap on the side you want him to go and moving him into regions where he is less dangerous and has poor angles of attack.

How far ahead?

How can this information be used to get a pay-off? That remains to be demonstrated, but let us return to our 'passing' question. Figure 2.20 illustrates a possible situation in which a mid-field player M is about to pass the ball to a running forward F. If the ball in a long careful pass takes 2 seconds to travel 40 metres, incidentally rising to a height of 5 metres, then the forward F running in at 7 m s^{-1} will only collect it in his stride if it is aimed at P about 14 metres ahead of him. This is much farther than most people would guess—although everybody knows that you pass ahead of a player few of us are sufficiently aware of the relative speeds to make a good guess at the best angle of passing— observation and calculation may help to improve football skill in this sort of way.

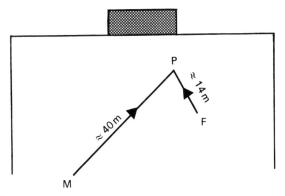

FIGURE 2.20

Goal kicks

Goal kicks raise another question—pass it out or belt it up the field? The latter gives a fair chance, presumably about 0.5, of losing possession, so why do goalkeepers do it? We must presumably compare this hazard with the probability of losing possession by a mistake in the chain of 'safe' passes up to the same mid-field position. Crudely, if the probability of success in each pass is the same, P_1, then for n passes

$$P_n = (P_1)^n.$$

The failure probability is then $1 - P_n$. For five passes, $P_n = 0.5$ requires $P_1 = 0.87$, which needs rather reliable passing.

At the higher levels of football even the long kick has a better-than-random chance of retaining possession. The forwards can give the goalkeeper a larger target to aim for by bunching together, while he can drop the kick a little short on their 'safe' side. There is room for observation, calculation and experiment here too.

The third dimension

In most of the discussion so far we have assumed that the ball travels in a straight line more or less horizontally, and much football is played in this way; however, the movement of the ball, if not the players, is in three dimensions and gravity as well as spin makes the flight curved. How far is this exploited and what other possibilities are there?

The long kicks exploit the parabolic trajectory—the cross or the corner kick normally aims, or hopes, to drop at the head of the striker, but often who gets it is fairly random with tall defenders dominating in many cases unless surprise is contrived. Otherwise the third dimension is used rarely, but often memorably. One of Scotland's moments of glory in the 1978 World Cup was when, against Holland, Gemmill broke clear at the edge of the penalty area. The goalkeeper came out and, 'spreading' himself, beautifully masked the goal as in Figure 2.19. However, Gemmill went beyond the straight-line model and, instead of hitting the ball hard, lobbed it over the goalkeeper and into the net. Equally, the coverage of a defender against passes past him only extends as high as his head so a lob can easily beat him. Why is this not used more?

There are various factors against it. It is slower—the ball takes a second to go up to 5 metres and another second to fall back to Earth. Its speed and angle must be better judged—a straight-line pass can be hit at any speed fast enough to beat the defenders, but a lob hit too hard will pass over the target player. Thus the error rate will be higher so the chance of completing a sequence of n lobs will decrease rapidly with n. This perhaps explains why it is used for decisive final passes aimed at penetrating the defence, rather than in the previous build-up of the movement. One still wonders if these skills could not be further developed and exploited—just as an occasional tennis player is astonishingly effective with a lob, in perhaps more difficult circumstances. Top spin helps there too!

Why does defence dominate over attack?

Why is the defence so dominant in football? Possession hardly ever leads to a goal. Various factors, such as error rate, the goalkeeper's hands and the off-side rule, may be important. How can one pursue this matter further? A detailed observation of televised football matches with notes as to why each individual move breaks down will provide the essential data for discussing these matters. Its collection and analysis, and the critical evaluation of various hypotheses can provide a valuable modelling exercise.

It is a historical fact that at each stage of development of the game, teams that emphasized defence have always tended to win—originally football was played with nine forwards and two defence. Probably the main reason is 'space'. A single attacker in the penalty area with a sight of goal and time to shoot will often score, while crowding sharply reduces the chance of such opportunities occurring and a greater number of forwards (and only one ball) does not increase it nearly as much.

Next we consider a more active approach.

Changing the rules

The effect of changing the rules provides a fertile ground for modelling of a quite creative kind, particularly if the P.E. staff can become interested. Five-a-side football in the gym or the games period allows short experimental games with altered rules. One could try, for example:

1 no goalkeeper
2 no off-side rule
3 one forward exempt from the off-side rule
4 the use of hands, but not holding the ball, allowed
5 no passing across any of the lines marked on the field.

or many other possible variations. The changes and the expected result should be thought out beforehand and the model predictions validated on the field. The difficulties of devising and carrying through reliable experiments, particularly in the human sciences, will soon emerge but fun should be had by most, if not all. Detailed record keeping is essential and valuable. Finding a satisfactory alternative set of rules needs more subtle changes than the above, such as changing the number of players—it could well fill some lunch hours.

Changing the game

There are those who do not enjoy such athletic extravagances as football but who may also, by this time, be quite fascinated by its analysis. For them an alternative challenge is to devise a board game that, reasonably accurately, simulates a game of football. Dice can provide the random element that reflects mistakes, with chances of success depending on the ambition of the move, e.g. 5/6 for a penalty, 1/216 (3 successive sixes) for a shot from the defensive half of the field. Rules must be realistic but not too complicated—a challenging problem.

Athletics

The running, jumping and throwing events of the athletics track have a real interest for some children. At a curiosity level they may discover and compare the average speeds of the runner over various distances. How does the four-minute miler compare to the sprinter? This sort of thing can be learnt from the record books. But what about the final sprint in a longer race—how fast are they going then? An estimate can be found with a watch and a careful look at a televised race. The last lap time is interesting, but not really good enough because the sprint begins later.

As usual a closer look can have some pay-off for those really involved with the situation. A sprinter, for example, may like to compare the number (and thus the average length) of strides of different competitors. Where in the race are the crucial 'gains' made? Careful timings on film can show how position, velocity and acceleration vary during a race, making clear what distinguishes the very good sprinter from the merely good. Is it who 'dies' last?

Throwing events tend to require sixth form mathematics, though on page 160 we give an explicit formula that can be investigated using a calculator to discover the importance of speed, angle and height of launching in fixing the distance a shot travels. Although the derivation of such a formula is beyond the capacity of almost all pupils at school, it can nevertheless be used by many of them as a basis for an investigation into throwing events.

Jumping events also concern the motion of projectiles—in this case the human body. Any complete discussion would be complicated but certain basic questions are intriguing. How does the good high jumper clear a bar at over 2 metres? A pole vaulter can clear over 5 metres. By jumping? Get several pupils to try a standing jump, touching the wall as high as they can—even the most athletic of them will hardly jump more than 0.5 m. Why is there this difference?

Someone may realize that their centre of gravity starts at about 1 metre above the ground, but this still leaves a problem. The energy of the run-up comes into it; indeed a film will show that the pole vaulter runs about twice as fast as a high jumper giving four times the kinetic energy and height. It is more difficult to see why the one can convert so much more energy than the other.

Athletics produces other fascinations. The way in which world record performances have changed in time contains surprises. While in the events, like sprinting, that depend mainly on natural endowment the records are now changing very slowly over the years, in some of the technical events the pace is actually accelerating. The effect of introducing new factors like the fibreglass pole in pole vaulting, and aerodynamic javelins can be seen. The stimulant effect of Olympic competition is clearly visible too; the world's best performances in post-Olympic years often show a clear drop in standards. The figures on which to base such studies are published and readily available.

One final tractable problem—to design the 'staggered' start for a track race where the runners must stay in their lanes either for the whole race, or for part of it, which is harder. Measurement (for the least able) and algebra (for the most able) both yield useful results.

2.4   People*

Personal relationships interest everybody and loom large in the teens. They usually seem to provide about half the problems produced by a class for a modelling lesson; some of the more practical have been discussed in Section 2.2. Surely, you may say, this above all is not mathematics. It is true that at first mathematics seems to have no place in understanding such qualitative human phenomena but we hope that, in reading this section, you will become convinced that it has something

* This section was written with Sheila Bryant.

to contribute to their better understanding, as one of a number of useful tools of analysis. The models we present have, of course, no claim to authority; they are a more coherent version of the sort of models that children, and adults, produce in fragmentary form in conversations about these issues, and which affect their actions.

What about discussing such problems in the classroom? Since these are areas of real concern to children we should like to be of help, but there are obvious dangers. Should the more sensitive areas be left well alone, or left to the teacher of civics, or religious or moral education? Perhaps—certainly they need to be handled sensitively and sympatheti-cally because any particular problem may be a source of acute anxiety to some child in the class. Teachers who already find that children talk to them about personal matters will probably know where the risks lie and venture cautiously, while others should probably stick to the more practical aspects of life. The make-up of modelling groups, and their choice of problems, allows a good deal of flexibility.

On the positive side, there can be a significant pay-off from discussing personal problems, in various ways. Some topics (no friends, smelly feet) have an inherent tension for those who suffer them; an impersonal but sympathetic analysis can help to reduce this. Some people feel that they alone have a problem which is in fact universal (like fear on entering a room full of strangers); the recognition of this reduces the problem. Most importantly, rational analysis can reveal the possibility of departing from instinctive or conditioned behaviour in some respects and thus can give hope by suggesting alternative approaches that may prove more satisfactory; this, of course, is the positive approach to problems that we have been advocating through-out. The main danger is that particular children will be made more anxious about their problems, so the teacher must be sure that the individual is not obviously identified and that discussion is sympathetic and supportive.

Humour

The nature of humour is an ancient puzzle—what causes people to laugh? Surprise, incongruity, shock, are all part of it but in what proportions? A study of the ingredients in a collection of, say, Irish jokes (also told in Holland as Belgian jokes, and in Belgium as...) may lead to amusing insights. Classify and count.

Integrating the handicapped

The Warnock Report to the Government recommends that, as far as possible, handicapped children should be taught together with others in normal schools. This will clearly help in decreasing the isolation of the handicapped, and in making other children aware of the difficulties of other people. It will, however, raise a lot of practical problems, and perhaps some deeper ones too, in organizing the life of the school. Working these out in detail, within limits of money and other resources, may provide an interesting session or two. Access to various places, ramps, wheelchairs, doors, toilets, space in the classroom, rotas of helpers may all enter. Handicapped children may be made more aware of their being different in direct comparison with the children around them (though they are probably well aware already); events should be planned in such a way that the children with difficulties are always included as far as possible.

Friends

Friends are of importance to everybody but they play different roles for different people. Some people seem to thrive on a fairly lone existence, while others spend a lot of time with their many friends. One way of looking at the differences in patterns of friendship is through a graphical representation, sometimes called a sociogram (see Figure 2.21). It works like this. Every child in the class is represented by a point. Each pair of friends is joined by a line; friends outside the class can be represented by short lines at the point.

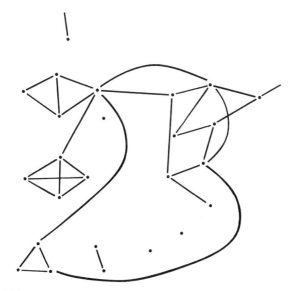

FIGURE 2.21

It is found that different children have very different patterns. Obviously some will have more friends than others within or without the class. Girls often tend to form tight groups hardly connected to the rest of the class but boys are more generally and loosely connected. Since sociograms identify explicitly the isolation of some children they may provide useful information for the teacher, but should *not* be open to the class.

There is no established way to classify friends as to how close they are, but some such scheme is useful if one wants to take the question further. There is something to be said for the scale used by the *Guide Michelin* to classify tourist attractions:

***   worth a special journey

**   worth a detour

 *   worth stopping if you are passing,

to which one should add:

A   acquaintance, with whom one interacts at work or otherwise in the course of life

O   the outside world of strangers.

It is worth debating whether a person has a single **** friend of the opposite sex, closer than any other, as would be expected by the standard model of marriage for example, but on most observational or analytical views of closeness this does not seem to be true, at least among teenagers. In looking for evidence you might consider:

– time spent together in conversation

– length of friendship

– apparent pleasure derived.

Again a diagrammatic approach will yield interesting results. Ask each person to put themselves at the centre of a piece of paper and

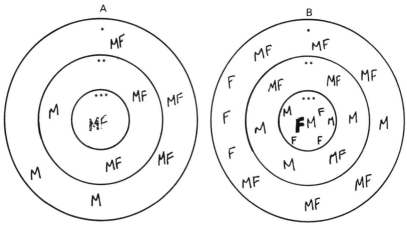

FIGURE 2.22

represent their individual friends by points whose relative closeness is indicated by distance out (closer means closer). One, two and three star circles can be indicated. The sex of each friend should be identified, but names need not be given. Two examples are shown in Figure 2.22. This is strictly a game for volunteers. There is no suggestion that there is a 'best' pattern, or that with friends 'more' means 'better'.

Making friends

In any modelling group the nature of friendship and the process of making friends seems to be an endless source of fascination. After the opening ideas session the blackboard may well look something like Figure 2.23. We will discuss these headings briefly to show a line of attack and pursue a few of them in more detail.

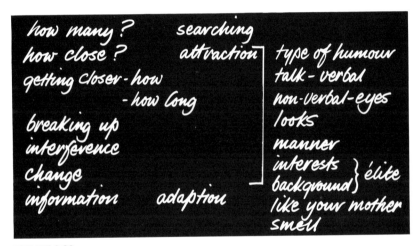

FIGURE 2.23

We have already said something about patterns of friendship which relate to 'how many?' and 'how close?'. Such a classification can clearly be improved but something simple may be enough, at least to begin with. The whole situation has so many facets that it is wise to stick to simple qualitative models at a first look.

Now let us look at the processes of finding (and of losing) friends. The group will be able to identify various phases in the process, provided they are given a definite context to think about—their own age and stage is easiest but they may have comments about earlier stages of their life and even ideas about mum and dad's generation. *Interpersonal attraction* by E. Berscheid and E. C. Walster (Addison Wesley, 1969) describes experimental evidence and other models of these phenomena.

Searching

The initial searching process in which we find potential friends from the enormous number of people we could meet is interesting in several ways. What is the nature of the search? What qualities are we looking for? How do we get the information? How long does it take? How many people do we look at?

Many people believe that finding friends, and partners, is essentially a search process but most of us seem to look at surprisingly few people in that search. Estimating how many is a nice problem. It is simplest if one avoids worrying about levels of friendship at the same time. With married people this can be done by asking the question:

'To how many people did you give some chance of marrying you, by getting to know them well enough to decide if you might be interested?'

This seems to allow people to make useful estimates—answers tend to lie in the 5 to 100 region with 20 as a fairly typical response. It is worth comparing the results of various ways of making the estimate: listing past associations, estimating how frequently one met 'interesting' people and over what period of one's life, looking at the rate at which one met new people and estimating a proportion of those likely to be interesting, etc. However estimated, the result usually strikes people as surprisingly small, considering the apparently eligible population, whether you think it is 25 000 000 or 2 000 000 000; it does seem that we are still nearer to the arranged marriage of Indian society than to a romantic search for the ideal.

This estimate may be quite unreliable, perhaps we are unconsciously scanning people and carefully picking out those we regard as potential friends before we recognize any wish to approach them. If so, and there is evidence of this, it cannot be on the basis of any very profound aspects of their character.

A similar discussion with a teenage group would relate to their current close friends of the same, or of the opposite sex. Again interesting patterns will emerge—the stability of the former, relative to the latter, relationships perhaps, the changes from 13 to 16, or from one generation to another.

When a model yields a surprising result the proper first reaction is to doubt the model. If it withstands further scrutiny however, as this seems to, the next step is to go on and look for underlying causes of the phenomenon. We will talk about one aspect in more detail later—the nature of *attraction*. Now for a more mathematical look at the search process.

First let's think about the question of how long we spend getting to know a given new friend. Is the sixteen-year-old girl who stays 'faithful' to her boyfriend for a year on a sounder track than the thirteen-year-old who changes hers every three weeks? How can mathematics possibly help in this situation. Well it can help you to recognize the conflict between getting to know somebody well, which takes a lot of time, and getting to know a wide range of people, which doesn't allow much time for each.

The second statement is easier to express mathematically. Let's define some variables and make a simple model.

Time with one boyfriend t weeks
(an average clearly)
Number of boyfriends per year n

Then clearly

$$nt = 52$$

$$n(t) = \frac{52}{t}.$$

The other aspect is more difficult. Let U be a measure of the understanding reached in time t. What can we say about $U(t)$? Perhaps:
- it increases steadily
- it takes some time before you begin to learn much
- after a long time you 'settle down' and don't learn much more

all of which suggests a graph like Figure 2.24.

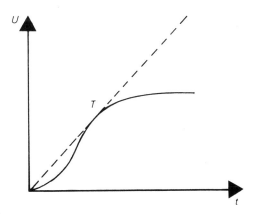

FIGURE 2.24

Now the 'total amount of understanding', call it experience E, gained in a year might be taken as

$$E = nU$$

which would then have a shape like Figure 2.25. This sort of balance between a rising factor $U(t)$ and a falling factor $n(t)$ giving a maximum is a very general property which is expressed rather clearly mathematically.

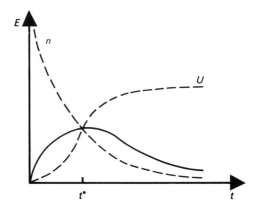

FIGURE 2.25

Note that it must be a product $n \times U$; $n + U$ won't do. Also it does depend on $n(0) \times U(0)$ being zero or some small value. It is also instructive to plot various forms of $U(t)$ and the corresponding $E(t)$, in order to gain some further appreciation of what various algebraic expressions really mean.

The way that rising and falling factors can combine to give a maximum is an important idea to learn. Although the algebra is a bit sophisticated, several numerical examples worked through with a calculator, plotted and interpreted will show quite clearly how it works. In fact, the maximum understanding per unit time is gained at the tangent point T in Figure 2.24—this idea is used in many problems involving the division of forces. Of course, to make further progress you want a more quantitative understanding of $U(t)$. Its absolute value is quite arbitrary (you can even choose a nice name for a unit—a sympa perhaps) but you can probably get some view as to how it depends on time. A quick linear gain of superficial reactions in the first few minutes conversation and then more slowly—if that is what you value then your t^* will be short. If, however, you recognize that the world is full of pleasant people, but those to whom you can relate closely in interest and values are rarer, then the initial conversations are merely a filter and serious discussion takes longer—weeks of spare time before you really begin to understand somebody. Later you may come to recognize there is a process of actual modification and harmonization of attitudes which takes many months, or even years. Whatever your sense of value there will be an appropriate $U(t)$ for you; it increases self-awareness to try to decide what it is.

Computer dating

Computer dating bureaux started in the 1960s. Initially they were regarded as something of a joke but the fact that they remain in business shows that they fulfil a need in providing at least hope, and possibly satisfaction in the search process we've been talking about. The rationale is simple—even if your t^* is only a week, you can still only discover 250 people in a five year period [not a large proportion ($\approx 10^{-5}$) of those of the opposite sex in this country]. Of course, they are not chosen at random, but the initial filtering system ('she seems nice') appears pretty crude and the numbers you even meet are quite small. (You might try to estimate how many people you meet in a year whom you could ask to go out for an evening—at various levels of brashness.) So some system that would match people on the important variables for compatibility and put them in touch may be of value. The computer dating bureau operates such a service among its clients.

Two aspects of the problem emerge: what are the important variables and how could you match them?

A class of teenagers will produce a list of variables very readily (education is never a one-way process). They may include:

physical attributes—size, eye, hair and skin colour, 'looks', voice
responses—quick, cheerful, active, talkative
habits—smoking, drinking, 'eating Chinese'
interests—pop, dancing, singing, sport, sex.

The matching problem is more subtle, mainly because of the choice between similar or complementary attributes. Does it work better to match 'cheerful' with 'cheerful' or with 'morose'; or do you, perhaps, need mainly similar but a few complementary attributes? Again views will not be in short supply. Devising a numerical measure of 'match' is more taxing but something should emerge. It may need to be worked out on a series of specific cases, which the class can provide.

The problem of organizing all this (and running a practical experiment) is highly educative, though care may be needed to avoid hurt feelings. It involves:

1 questionnaire construction (see Section 4.4)
2 the devising of a multiple-choice format and the layout of the answers on punched cards
3 the devising and flowcharting of an algorithm to do the matching job on a simple scale
4 if possible, the writing of a computer program to carry this out
5 estimating the processing time required for the class, the school year group, the school or the whole city.

All these activities are valuable applied mathematics. They may also be of value in bringing out into open discussion, in a fairly neutral and practical emotional environment, matters which for some children have an aura of tension. For the child who does not find it easy to make friends the obvious fact that the chance of finding a close match in a class of 30 is relatively small, while a larger arena offers a much greater probability may be a new and comforting discovery.

Phases of friendship

Qualitative models can be constructed to describe the phases through which friendships may go. Discussion of *attraction* between individuals will usually lead to the conclusion that attraction is a function of the pair and that single person characteristics such as appearance and interests are of less importance than the matching between the pair. A model of the path of a friendship may involve the identification of phases such as *search* (looking for positive attraction), *investigation* (looking for negative attributes in the hope of their absence), *confirmation* (positive enjoyment of relationship) and *absorption* (each taking the other as an integral part of the environment). You will get a lot of introspective comment on the validity of such a model—it is valuable but difficult to look for other means of testing it. *Adaption*, the process which enables people to fit and work together by minimizing areas of stress is worth discussing, as is the process of *decay* of a friendship and the way one friendship may *interfere* with another.

Making the most of a small bedroom

A complete change of subject takes us to a postscript on people. A group of four pupils acting as consultants to a boy who wanted help in planning his bedroom, produced an ingenious set of designs aimed at saving space and promoting comfort (one example is shown in Figure 2.26). Problems of design and planning develop a wide range of modelling skills; indeed courses on metal work, design, and technology may well provide the best ground for real problem solving in the current curriculum. We believe that they should be encouraged and expanded to take wider objectives.

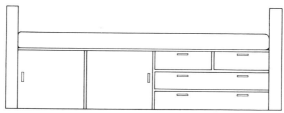

BED WITH CUPBOARDS AND/OR DRAWERS UNDERNEATH

FIGURE 2.26

3

TACKLING REAL PROBLEMS IN THE CLASSROOM

While people have been tackling real problems in a more or less mathematical way since the beginning of time, the explicit teaching of these skills is only just beginning to be explored. We must expect it to take some time to learn even moderately effective methods of teaching modelling, and much longer really to optimize them. Nonetheless, it has already been shown that many teachers can achieve satisfying, sometimes even dramatic results using a number of significantly different approaches. Further, it has been possible to identify a number of key variables in the teaching of modelling and so to give the interested teacher a framework within which to make his own explorations.

This chapter gives an account of both these aspects. It describes some of the most highly-developed projects worldwide. It identifies important questions that still have to be explored and summarizes the few important results that reasonably systematic research has so far yielded.

Teaching effective ways of tackling real problems is still a pioneering area with plenty of room for inspired intuition, but such enterprise is more likely to be successful when it is based on an awareness of what has been done.

3.1 Action mathematics in action

While the teaching of skills in modelling realistic problem situations is in its infancy, it may be helpful to describe some of the most vigorous offspring. Experiments have been going on in the last ten years or so in various places around the world with students of various ages (we have already referred to some of them). They vary in their amount of emphasis on model formulation, in the realism of the problem situations, and in their scale. There has been more activity in the primary school and university age ranges than in secondary school, probably because of the difficulties created by subject boundaries and by examination pressures in that sector.

The outstanding project is USMES—Unified Sciences and Mathematics in Elementary Schools—based in Boston and directed by Earle Lomon of MIT. It has a well-developed programme, which in 1978 was being used by 2 000 teachers and 60 000 children in 500 primary schools all over the United States and elsewhere; the project is also now being used in about 15 high schools without very great revision. We shall review this approach in some detail below. In this country there is as yet no comparable scheme, but a number of enterprises have related aims. Some of the syllabus reform projects of the 1960s aimed to emphasize the useful applicability of mathematics—the Mathematics for Education in Industry (MEI) project was one example. However, the major objectives

"It makes new demands on the teacher"

remained the teaching of mathematical technique, and applications were confined to standard models, usually of highly-idealized situations.

An interesting exception is the *Mathematics Applicable* series produced by the Schools Council Sixth Form Mathematics Project, directed by Christopher Ormell and published by Heinemann. This fully developed one or two year A/O Level course is centred around the modelling process. The initial objective is to gain a thorough understanding of various mathematical models (e.g. linear functions, exponentials, probability, calculus, etc.) aided by the students' understanding of a rich variety of concrete situations which they model. There is a particular emphasis in this project on developing translation skills. Further details on *Mathematics Applicable* are given in Section 4.1, page 119.

Other experiments have been mainly at university level. The author has been developing experimental courses on the modelling of realistic situations for undergraduates and extramural students over more than a decade—teaching methods will be discussed in the next chapter. R. R. McLone of Southampton University has developed a set of courses on modelling for undergraduates. More recently the Open University has revised its mathematics foundation course and Block V of the new course M101 is on mathematical modelling, including standard models and realistic new situations with a balanced emphasis on the whole modelling process; their new course for teachers is described on page 88. There are ideas which can usefully be borrowed from all these areas for use in the

11–16 age range but, as we have stressed above, nobody has established material at that level.

USMES in action

We now give a brief review of the work of this project, mainly by quotation from its own introductory material. Although few educational approaches transplant directly from one country to another, any teacher interested in realistic applications of mathematics will learn from the USMES material. To give some flavour of the work we give first a brief overview of the project, followed by the introduction to a particular 'challenge', Getting There, which asks the class to

'Find ways to overcome the difficulties in getting from one place to another.'

This is taken from the Teacher Resource Book, which gives a detailed discussion of the problem and children's responses to it; there is one for each challenge.

USMES (Unified Sciences and Mathematics for Elementary Schools) is an interdisciplinary program, developed under grants from the National Science Foundation, that challenges elementary school students to solve real problems from their school and community environment. Currently, the program is being used in suburban, rural, and inner-city schools located in 32 states and the District of Columbia.

Many of my fifth-grade students became disgruntled with their playground activities. Consequently, when the complaining began, I challenged them to do something about it.*

The curriculum is organized into 26 units based on real problems stated as 'challenges'. All have been developed in the classroom by teachers and students. Most units can be used in grades K through 8, and they can be introduced to students in any order. The level at which children approach a challenge, the investigations they organize and conduct, and the solutions they find vary according to age, ability, and interest.

The students' first response was to order some new playground equipment. But after looking at prices, we realized the problem couldn't be solved in this manner; we had only $150 of district money with which to work. It was at this point that my youngsters suggested making some equipment ourselves.

Children working on USMES units tackle real problems such as a busy intersection near the school, classroom furniture that doesn't fit the students, or playgrounds that are crowded or uninteresting. Each problem is 'real' in that it has an immediate, practical impact on students; there is no 'right' solution; it requires students to generate ideas and procedures for finding solutions; it can be resolved by students; and it requires several stages of class activity, from identifying a problem to collecting and analyzing data and acting on the results.

Our initial brainstorming session dealt with 'Things to Consider' or 'Problems to be Faced'; for example, what new equipment would we like to have? Who, besides our class, would use the equipment? How big should the equipment be? What kinds of measurements do we need to make? This session gave us the direction we were to follow during the ensuing weeks.

Solving real problems is an interdisciplinary activity requiring the use of skills; processes, and concepts from science, mathematics, social science, and language arts. Accepting a challenge, students become involved in defining the problem; conducting observations; taking measurements; collecting data; analyzing data; using charts, graphs, maps, and statistics; discussing ideas, procedures, and solutions; developing and clarifying values; making decisions; learning to work productively in small groups; and communicating findings to others.

The first problem the students wanted to discuss was how large we should make the equipment. They decided to measure everybody who might use the finished products, and determined that they should collect information on height, weight, hip width, arm length, and leg length.

Another question was the size of the playground. Two boys spied the USMES 'How To' cards on measuring large distances. Excitedly acting on an idea contained in the cards, each constructed a trundle wheel for measuring the school grounds... When they began drawing the playground and the existing equipment, the children gained a great deal of practice in division and multiplication. As the need for additional skills surfaced, we held skill sessions on using ratios, proportions, fractions, scale drawings, and indirect measurements.

The students designed and later constructed an adjustable 'battle horse'; a double see-saw; two swinging trapezes, one for younger children and one for older children; a swing; and an obstacle course. They no longer complain that English, math, science, or social science is boring and meaningless. For the first time, they can see that skills in all these subjects are essential to everyday living.

* Quotes taken from the log of USMES teacher, James E. Brown, Shapleigh, Maine.

This extract gives an impression of the sort of activities involved in the USMES approach, and shows how far it shares the aims we have been putting forward in this book. The teacher has a variety of resources at his disposal. The 'How to' cards and booklets help to teach the pupils skills necessary for tackling the challenge in which they are engaged. The Design Lab is a collection of tools and supplies the children can use for design and construction of useful models and apparatus; the *Design Lab Manual* discusses the problems in setting up and running Design Labs of various levels of sophistication. The *USMES Guide* also describes how work on USMES can be linked with other school activities and lessons. Most important of all, the Teacher Resource Book for each challenge describes in considerable detail the nature of the challenge, the various aspects of it and possible approaches to it, including essential background information and references for further reading; it includes extensive quotations from the logs of teachers who have worked on the challenge and examples of children's work. All these resources and the USMES approach are described in much more detail in *The USMES Guide*, which is essential reading for any teacher interested in trying these things out in practice. The material may be obtained from the publishers, ·Moore Publishing Company Limited, Box 3036, Durham, North Carolina 27705, USA, from whom price lists can be obtained.

We now give extracts from the introductory section of the Teacher Resource Book of one particular challenge as a useful way of giving a better 'feeling' for what is likely to be involved; we regret that we do not have space for the extensive quotations from the reports of teachers and pupils which these resource books contain.

Challenge:
Find ways to overcome the difficulties in
getting from one place to another.

Possible Class Challenges:
What can we do to help students find their
way around the school?
What is the best way for us to get to the
neighborhood library, playground, etc.,
from school?

Children in a variety of grade levels may respond enthusiastically to this challenge as they discuss the problems they face in getting around their school or community. In some classes the focus of the children's investigations is getting around the school or simply going from their homes to school and back, while other classes may investigate ways to get to and return from recreational, educational, and cultural facilities within the community. Once students have realized that a problem of getting from one place to another exists, either for themselves or for others in the school and/or community, the children may decide to work in small groups on different aspects of the problem. The flow chart suggests some activities that may take place in the class.

This challenge has been successfully tried at grade levels from kindergarten to six. It must be emphasized that a sufficient background for USMES challenges is much less than that required in the more traditional educational context in which the student is expected to proceed rapidly along a predetermined route. Once challenged, the student may acquire the skills and concepts when needed in his/her search for some solution to the challenge. The following may be said about prerequisite skills for this unit:

1 Counting—many children learn to count by rote; USMES provides the children with an excellent opportunity to learn and understand counting.

2 Graphing—graphing skills may be learned as the need for them arises.

3 Measurement—measurement skills may be learned as each new activity is begun and improved when additional or new data is required.

4 Division—the calculation of averages is not necessary, sets of data can be compared graphically and by subtracting medians (half-way values) and ranges. To make scale drawings, young children can convert their measurements to 'blocks' on graph paper.

2   CLASSROOM STRATEGY FOR GETTING THERE

A teacher preparing to introduce the Getting There challenge will find it useful to become familiar with USMES written materials including the USMES Guide, the Getting There Teacher Resource Book, those Background Papers pertinent to the unit, and the Design Lab Manual. In this way he/she will be better able to anticipate and deal with the various situations that may arise in the course of the class's investigations of the challenge.

The children's commitment to finding a solution to the challenge, 'Find ways to overcome difficulties in getting from one place to another,' holds the key for success of the Getting There unit in a class. Because the challenge provides the focus for student activities, the introduction of the challenge is an extremely important part of the USMES strategy. However, this challenge can and should be restated to fit the particular situation, especially if the restatement makes more sense to the children. For example, fifth-grade students in one school were challenged to find ways to help students find their way around the school.

One way in which the Getting There challenge can be introduced quite naturally is during a spontaneous discussion of some recent related event, such as when students want to get to a particular room in the school and are unable to find their way.

Children in a kindergarten class began their investigations of the Getting There challenge after a discussion of the class's trips to the school library. The children went to the library every other week but really did not understand where it was or how they got there. After further discussion, the whole class tried to find the library and gave directions which their teacher recorded. When they were unable to get there, the children met for a group discussion to plan what to do.

Once the challenge has been introduced and the class has determined that the problem does exist for them, possible solutions are suggested, acted on, and tested. It is important that the students assign priorities to the various tasks that must be performed so that some groups do not become stalled in their progress because others have not completed their tasks. The students then form groups to attack different aspects of the problem. However, the teacher may find that having too many groups of only two or three students attacking different aspects of the problem at the same time not only makes it difficult to be aware of the progress or problems of each group but also makes it difficult for thorough investigations of the challenge to be performed by the class; the larger the number of groups, the smaller the membership of each group, thus lessening the chance for varied student input and interaction.

As initial work is completed, the students regroup to investigate other parts of the problem.

> *Fifth-grade students in one class working on the Getting There challenge of determining ways to help students find their way around the school, met frequently to review the challenge and their work on the unit. After the class completed investigations to prove that the problem did exist and met to brainstorm possible solutions, they formed several small groups to work on areas designated by the class, for example, wall signs, 'traffic man', room numbers, and map rack. As the unit progressed and as discussion showed that additional data or tasks were necessary, the students regrouped to complete the necessary work.*

The teacher also needs to be aware of the many activities taking place in order to ask questions and to provide resources as needed. Because each group should be allowed adequate time to report to the class on the procedures they used and the results they obtained, having too many small groups will result in too little discussion time for each group.

Open-ended questions asked at appropriate times by the teacher serve to help the children refocus their attention on the challenge and stimulate the students' thinking so that they will make more extensive or comprehensive investigations or analyses of data. Examples of such questions can be found later in this section.

During class work on an USMES challenge, teachers serve as a resource that children can consult when in need of information and suggestions on where to locate it. In the course of their investigations there are times when students become stalled in their work and are in need of specific skills, e.g., making graphs, finding the median, designing a survey. Teachers may take these opportunities to hold skill sessions either with groups of students or with the entire class. Sets of 'How To' Cards, such as 'How to Make a Bar Graph Picture of Your Data', may also be used by students as the need for specific skills arises.

4   FLOW CHART

On the following page is presented a flow chart of various possible activities that can occur during the investigation of the Getting There challenge. The chart indicates some of the discussions, observations, calculations, graphing, and constructions in which the students may become involved during the course of their investigations. It is not intended to be a complete list of activities and investigations can and will occur as students suggest other ways to approach the challenge.

Some of the other activities show how comprehensive investigations can evolve from the students' initial ideas. Thus, class discussions of the challenge and its possible solutions lead to many varied activities. However, a class may perform only some of the activities indicated, or it may follow an entirely different order, as well as different activities. This chart is designed merely to indicate one possible progression of student activities in response to the challenge from its introduction to its solution and to the beginnings of other possible investigations or other USMES challenges.

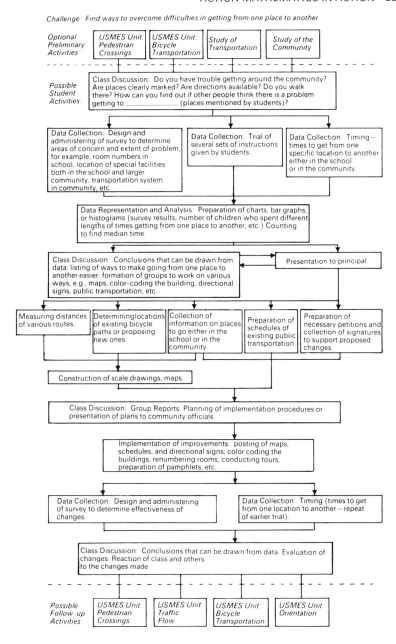

Challenge: Find ways to overcome difficulties in getting from one place to another

Optional Preliminary Activities:	USMES Unit: Pedestrian Crossings	USMES Unit: Bicycle Transportation	Study of Transportation	Study of the Community

Possible Student Activities:

Class Discussion: Do you have trouble getting around the community? Are places clearly marked? Are directions available? Do you walk there? How can you find out if other people think there is a problem getting to _____ (places mentioned by students)?

Data Collection: Design and administering of survey to determine areas of concern and extent of problem, for example, room numbers in school, location of special facilities both in the school and larger community, transportation system in community, etc.

Data Collection: Trial of several sets of instructions given by students.

Data Collection: Timing—times to get from one specific location to another either in the school or in the community.

Data Representation and Analysis: Preparation of charts, bar graphs, or histograms (survey results, number of children who spent different lengths of times getting from one place to another, etc.) Counting to find median time.

Class Discussion: Conclusions that can be drawn from data; listing of ways to make going from one place to another easier; formation of groups to work on various ways, e.g., maps, color-coding the building, directional signs, public transportation, etc.

Presentation to principal.

Measuring distances of various routes.

Determining locations of existing bicycle paths or proposing new ones.

Collection of information on places to go either in the school or in the community.

Preparation of schedules of existing public transportation.

Preparation of necessary petitions and collection of signatures to support proposed changes.

Construction of scale drawings, maps.

Class Discussion: Group Reports. Planning of implementation procedures or presentation of plans to community officials.

Implementation of improvements: posting of maps, schedules, and directional signs; color coding the buildings; renumbering rooms; conducting tours; preparation of pamphlets, etc.

Data Collection: Design and administering of survey to determine effectiveness of changes.

Data Collection: Timing (times to get from one location to another—repeat of earlier trial).

Class Discussion: Conclusions that can be drawn from data. Evaluation of changes. Reaction of class and others to the changes made.

Possible Follow-up Activities:	USMES Unit: Pedestrian Crossings	USMES Unit: Traffic Flow	USMES Unit: Bicycle Transportation	USMES Unit: Orientation

We now give brief descriptions of some of the challenges the project has found can be profitably tackled by children in the age range 6–11; they seem to us to constitute a real challenge to our preconceptions as to what can be achieved in bridging the gap that so often exists between the school curriculum and everyday life.

Bicycle transportation—Find ways to make bicycle riding a safe and convenient way to travel

Classroom management—Develop and maintain a well-run classroom. (How can we make the class run more smoothly? How can we improve the daily class schedule? How can we keep the room neat and running smoothly?)

Describing people—Determine the best information to put in a description so that a person can be quickly and easily identified.

Designing for human proportions—Design or make changes in things that you use or wear so that they will be a good fit. (Design tables that would be comfortable for students in your class. Determine how many sizes of Design Lab smocks should be made for students in your school for comfort and reasonable cost.)

Eating in school—Promote changes that will make eating in school more enjoyable. (How can we improve the lunchroom environment? How can we improve the service in the lunchroom? How can we improve the food we have for lunch (snack)?)

Growing plants—Grow plants for _____. (Children determine the specific purpose, such as for gifts, for transplanting into a garden, for selling, etc.) (Grow plants to make the room or school area more attractive. Grow plants for sale or display at a school fair or carnival. Grow plants to feed animals for the school zoo unit. Grow plants and have them bloom by Mother's Day.)

Mass communication—Find a good way for us to tell many people about _____ (topic, problem). (Inform people at school and in the community about the problems of vandalism. How can we tell other people how we feel about school? Find the most effective mass communications method to teach the metric system to people in school.)

Pedestrian crossings—Recommend and try to have a change made that will improve the safety and convenience of a pedestrian crossing near the school.

Play area design and use—Promote changes which will improve the design or use of our school's play area. (Find ways to make our play area less crowded and safer. Make changes in equipment and game areas to make the play area better for the whole school.)

Protecting property—Find a good way to protect your _____ (property in desks or lockers, bikes, tools, animals, Design Lab tools, etc.) (How can we protect our bicycles? How can we keep things from being taken from the Design Lab? Design a good system that will prevent the loss of articles from our lockers.)

School rules—Find ways of influencing rules and the decision-making process in the school. (How can we remind children in the school to obey the school rules? How can we change the way school rules are made?)

Soft drink design—Invent a new soft drink that will be popular and can be produced at a low cost. (Invent a soft drink that could be sold in the school cafeteria. Invent and produce an inexpensive soft drink for a class party.)

We conclude our description of USMES with some brief remarks on the results of its evaluation in a substantial number and variety of independent surveys. For those who wish to know more, a summary of evaluation results, with references, is obtainable from the Educational Development Center, 55 Chapel Street, Newton, MA 02160, USA, which administered the project for the National Science Foundation.

From its beginning, USMES had among its aims the improvement of the higher-level cognitive strategies of problem solving and decision making of pupils, of the learning of basic observational, quantitative and communication skills, and of the attitudes of students towards the power of learning, towards their own ability to cope, and towards their role in improving society. The results of these surveys seem to show significant advantages in meeting all these aims in approaches based on real problem solving, and in the specific approach used by USMES in particular. For example, in one situational test, each student was given three note-books of differing size and number of pages, quality of paper and price, and asked to recommend the best for purchase in quantity by the school; pupils from USMES classes were clearly better in giving objectively measurable reasons for selecting a note-book and in the number of tests suggested or performed to substantiate the reasons given. In two other tests rather artificial problems were used and no significant differences were shown. Other tests indicate that this positive correlation of real problem solving ability with USMES training is at its greatest in those schools where the approach is pursued most consistently. There is also evidence of more effective learning of basic skills—here the difference between the USMES and the control group are all positive but fairly small, and must be regarded as encouraging rather than definitive. A study of pupil activities in USMES sessions indicates why this may be so—two-thirds of the sessions requiring the effective performance of some mathematical task, for example. The attitudes of pupils towards working with USMES showed that a very high proportion, in each case about 90 per cent, of pupils thought that USMES was good fun, important, and demanding. A similar proportion wanted to do more, in three surveys taken over a period of five years.

In all this evaluative work, encouraging as the results are, the quality and the coherence of the teaching seem to be an important factor. It confirms the impression, derived from many other curriculum development projects, that new ideas should be introduced slowly and carefully, with a sharp eye open to ensure that what occurs in the classroom is effective and in accord with the aims.

Real problem solving—the Open University teacher's course

In 1977 some work began in England that was largely inspired by the USMES project. A team at the Open University was formed to produce a course for teachers entitled 'Mathematics Across the Curriculum'. They were working against the background of a widely expressed concern that children were leaving school ill-equipped with the basic mathematical skills demanded by employment and adult life generally. The work of USMES suggested to the team that, by giving children a chance to tackle real problems and find their own solutions, basic mathematical skills could find some meaning for children. While 'skill-getting' facilitates 'skill-using', it may equally be true that skill-using can provide the kind of motivation for skill-getting that appeared to be sadly lacking.

Since the Mathematics Across the Curriculum course was aimed at teachers rather than children the team developed a slightly different approach to that of USMES. Instead of providing teachers with a series of tailor-made problem situations, they chose to explore the process of real problem solving and focus on the role that the teacher must play in order to make use of that process as a context for developing mathematical thinking. They decided that the teacher's role has three main aspects as creator of an environment, facilitator and intervener to stimulate thinking.

To understand the facilitator role, it was felt that teachers would need to know in some detail what might be expected to happen during a real problem project. With this in mind, the team developed a model of the process that they felt was simple enough for teachers to hold in their heads as well as comprehensive enough to cover most of the activities that the children would undertake. The result was the acronym PROBLEMS, which stands for:

Pose the problem
Refine into areas for investigation
Outline the questions to ask
Bring the right data home
Look for solutions
Establish recommendations
Make it happen
So what next?

The team supplemented it with a series of *self-organizing questions*, designed to suggest the kind of prompts that children might need if they were to achieve their goals satisfactorily. For example:

Outline the questions to ask:

1 What questions do we want to ask?
2 Would the answers help us solve the problem?
3 Would we be able to find answers?

It is the self-organizing questions that form the basis of the intervener role that the teacher has to play.

This brief description covers that part of the Mathematics Across the Curriculum course that is concerned with real problem solving. The process is described in much more detail in the units entitled Getting Started, Making Plans, Looking for Answers and Finishing Off. Before teachers study these units (which represent about 40 hours of study time), they attend a two-day residential school in which they have a chance to experience real problem solving for themselves. As they study the units, the teachers carry out a real problem project in their own classrooms. At the end, they are asked to evaluate what has happened and produce a write-up of the project that the children in their class chose, worked on and, it is hoped, solved to the satisfaction of everyone concerned.

Finally, how do teachers react to the course? As part of the course development, the team worked with a group of teachers in the Walsall area. Here is how the reactions of one teacher in the development group were summarized in *The Teacher* (13.7.79).

Mr Bob Poyser, the class teacher and deputy head of the school, had given his pupils the task of *trying to save the tuck shop* which the head teacher was planning to close down.

Group delegations were wandering around the school measuring up cupboards, other groups were huddled around calculators and maths graffiti had appeared all over the blackboard.

'At first we had a jumble of ideas about what we should do to justify its existence', said Mr Poyser, 'but these were sorted out, by the kids not me, into certain areas.

They are using graphs from surveys they made up themselves. They had to think up the right questions in the survey to get the right answers. They did a survey on pocket money and found out that most of them were only given 10p so that most things in the tuckshop would have to be low-price goods to make them sell.

They had to work out volume when it came to storage space.

When we started this project the kids were using fractions but had no idea of percentages, so I did a lesson on that, and most of them seemed to grasp the concept first time round.

We also had a lesson on using calculators, and topic webs are now no problem whatsoever.

The experience they have gained from trying to solve this problem has been very rich indeed, and they have wanted to do it and to learn because it is very real to them.'

One of the discoveries the children made for themselves was that many of them had used the tuckshop as a stopgap to dinnertime, as they did not have any breakfast before leaving for school.

Teachers speak

It would be wrong to give the impression that all tackling of real problems in the classroom was based on major projects. Some individual teachers can and have developed their own ways of achieving some of the same ends, stimulated by, and drawing ideas from, various sources including their own experience of using mathematics. The following comments by such teachers gives some flavour of the variety of possibilities. First, some first attempts:

1 J.M. writes, sending some of the handsome brochures of advice on various problems which the class of above-average 15-year-olds had produced: 'Well, here they are! I was pleased with them although it is difficult to judge whether they are the right sort of thing. We spent about five lessons in all and today I spent another lesson doing some of your problems. I started using the problem that a girl wanted to train to be a teacher and her parents were going abroad—write down all the factors worth considering. They were much better at actually writing down factors. We then did the potato problem*—they started but were still much too vague. I reproduced the argument on the board and although I lost some of them, I think they got the point about being specific.

   Not all the projects were totally successful.

Horse sense (c.f. page 40)
The group worked very hard collecting information but they copied too much out. The emphasis on a consultative document meant they wrote everything down but did not analyse enough.

How to get a horse from Plymouth to Peterborough
This started out really well with many suggestions for approach—visit B.R. and travel agents—these were not helpful and they were disappointed. Did try to involve 'maths', e.g. graph—very odd ideas. Not very good at expressing their ideas in words. Not accurate enough!

How to make the most of a small bedroom (see page 76)
Again started well but did not go much further than drawing furniture. The model of a bedroom could have been extended to other shapes.

The class said that they enjoyed the exercise and I hope will have gained something from it.'
The teacher is much more critical of the results than we were—several of the pieces of work were decisive and balanced.

---

* Are cheaper, smaller potatoes a better buy? The amount lost in peeling and the extra time used are important factors.

The worry that the mathematics used is much more elementary than that the children 'know' is always a problem until the gap between acquaintance and real mastery is accepted.

2 E.B. has been developing the use of 'problems from the outside world' for some time in a selective school where most of the children are of high ability. She presents problems which have at least one fairly well-defined question. The first example is straightforward.

'This one went down well and I think hinges on getting cut-price cornflakes.'

Train tickets

On Sunday five of us have to travel to London for the day. British Rail offer a variety of fares, and I want you to work out what the different fares will cost and so find which tickets we should buy. Three of the five travelling are children. We have got two free Kellogg's tickets and one token towards the third free ticket (two tokens per ticket and one token per packet of Cornflakes). Here are the facts:

Cheap Day Return	£7.05
Capital City Fare*	£4.80
Children travel at half price.	

The rules governing the use of Kellogg's tickets are: (1) each child must travel with an adult; (2) each adult can accompany two, but not more children; (3) the free tickets are not available with 'special' price fares marked*.

Kellogg's Cornflakes cost 32p.

But others are more open.

Moving house

An elderly couple are leaving a large house in Chilham, Kent, and are moving to a small cottage. They have given their three daughters J, P and E a lot of their excess furniture. J lives in Wales, 153 miles from London, P lives in London 56 miles from Chilham, E lives in Birmingham 110 miles from London.

The estimate from the removal firm for a round trip dropping P's furniture in London, E's in Birmingham, and J's in Wales is £163. How would you divide the cost of removal fairly between the three sisters?

'I gave this problem, along with lists of furniture, to an intelligent class of girls in their fourth year. I had taught them since they came to the school. They were used to the school, and they were familiar with my methods and my 'problems from the outside world'. In the initial discussion we decided that each daughter should pay according to the distance that she lived from Chilham. We also talked about the proportion of the cost that arose from loading and unloading, and the proportion that was from travelling round the country. The loading and unloading was to be weighted according to the quantity of furniture being delivered, but the travelling was to be a sum based solely on distance.

(Some solutions ignored this refinement, and multiplied the distance by the weight to find what proportion of the cost each should pay.)

After the general discussion the class settled down in groups or by themselves to produce a solution. The variation in the final answers was remarkably small, though the methods used and the weights given to the loads for each daughter varied considerably. Some worked in percentages throughout, others in fractions. I did not get a solution from every member of the class and those that had worked alone produced the best work though they were not necessarily the best mathematicians.

The final answers were, as usual with this sort of problem, presented to a ridiculous degree of accuracy, to the nearest penny, instead of the nearest £5.

In the end E and J paid £70 and P paid £23.'

3 Finally, P.T. writes:

'*Lesson* 1

A small third year class of 'low' ability (aiming for CSE 2/3) suggested these problems: getting work wrong, French, homework too hard, heights, falling off cliffs, losing boyfriend, getting old (?!), exams (4), dentist (2), dark streets, motorbike crash, maths, dad away on oil-rig, brother's accident. Only one had no problem to suggest.

Since exams were last June (and there were several with this worry), I decided that this could do with investigating. So far I've found no reason for this long continual preoccupation.

The points raised are shown in Figure 3.1, and as our lessons are rather short (35 minutes on this occasion) I asked them to consider the matter of revision for homework!

The school exam timetable involves several 'private study' sessions, when the person sitting next door may be writing an exam.

Lesson 2

The gap (1 week) has been too long. We've all lost our impetus. However, we/I persevered. The homework had produced only one

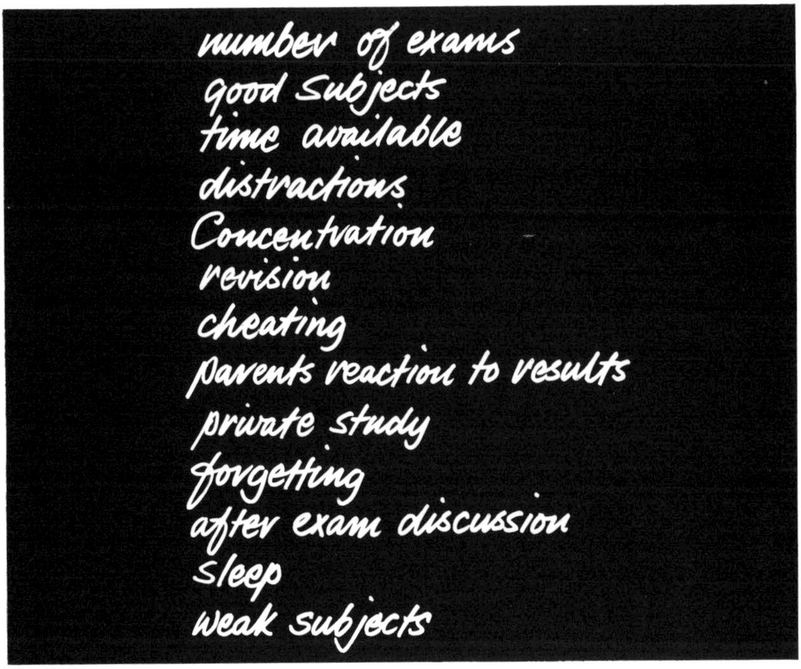

number of exams
good Subjects
time available
distractions
Concentration
revision
cheating
parents reaction to results
private study
forgetting
after exam discussion
sleep
weak subjects

FIGURE 3.1

girl who divided her time mathematically. The others had written out beautiful timetables and then allocated 1 or 2 hours to each subject mechanically.

So we looked at variables:

T = total time available
d = number of days until exams start
h = hours per day available
X = time per subject
S = number of subjects

and eventually got $X = T/S$; $T = d \times h$ so

$$X = \frac{d \times h}{S}. \tag{3.1}$$

The question of varying time according to difficulty of subject had been brought up earlier, but I put it off until stage (3.1) was reached.

We got completely bogged down in ratios. So I decided that I would profitably allocate an ordinary lesson to this during the next week, and then return to the attack.

For homework I set the task of writing instructions to enable *other* pupils to use (3.1). This was done quite well.

Lesson 3

We tackled ratio again—relating it to our problem. I felt that more complex ratios would overload them at this stage so we merely divided subjects into two groups: *B*ad and *G*ood. By allocating double time to *B* subjects we now had $S = G + 2B$ in the formula. During the rest of the lesson we tried different values of the variables and worked out results.

Homework was set to write up the work.

I felt that this was well done. These kids are obviously better at English than maths! (Presumably that is why they are with us.)

However, no one had used the *result* of the formula to state specifically how many hours would be given to *G* and *B* subjects. I investigated and found that most of them knew, although they hadn't written it down.

EPILOGUE (one week later)

Me   (thinks) "Thank goodness I've finished with 'problem' maths."

       (out loud) "Right—get out your books."

Class "What books? It's *PROBLEMS* today."

Me   "But I thought you didn't like them?"

Class "Oh *yes* we do!"'

Beginning to tackle real problems—a 'Starter Pack'

Many teachers are interested in helping their pupils to acquire modelling skills but, with reasonable scepticism and many other pressures on their own and their pupils' time, cannot make the sort of commitment that USMES or the Open University course imply. For some the ideas and examples in this book may be enough to enable them to explore the possibilities on a smaller scale but others will want a more specific and structured package, with lesson plans and pupil material. For them we have developed this 'Starter Pack' on modelling.

It aims to occupy 5–10 lessons, and in that time to cover the range of activities involved in real problem solving. The major difficulty most teachers find in such work is one of style, because open problems tend to demand an individualized pupil-centred investigative approach; the Pack meets this by including as a major element, structured modelling exercises, with marking schemes, which develop particular modelling skills which research has shown to be important in overall modelling ability. Pupils and teachers find these enjoyable and straightforward.

The package begins with a lesson which aims to show that mathematical analysis can be useful—the problem chosen is that of the child 'seeing the blackboard' over or around the heads of children in front. A preliminary class discussion will not usually make a lot of progress; the teacher is then intended to use the detailed notes provided to analyse the situation and its implications for classroom design. There follow several lessons of modelling exercises—on thinking of factors or variables, selecting and generating graphical relationships between variables, identifying specific key questions, estimating and modelling itself. Plenty of exercises are provided (the teacher and pupils are encouraged to produce more) and for many of them, marking schemes are supplied. This work also develops conventional mathematical skills, particularly in the use of graphs. There is then a gentle lead in to complete problems with carefully discussed solutions of 'how to choose between three sample notebooks for bulk purchase by the school', and of 'whether his parents should buy Terry a bike to travel to school', which may be used as a test problem. Further problems are also suggested.

The package is now undergoing further trials. Ultimately, it will probably be published but, at the moment, interested teachers may obtain it from The Shell Centre for Mathematical Education, The University, Nottingham, NG7 2RD, for the price of £1.20, post free.

PAMELA—a problem collection

PAMELA is a collection of Problems in Applied Mathematics from Everyday Life Applications. The problems in the collection all arise outside mathematics and, as usual, the process of modelling will begin and end there. They come from everyday life (or every-year life at least) and should be recognizable as problems by the interested, though perhaps mythical, man-in-the-street; this restriction is designed to rule out most of the myriad of problems arising from other school subjects or from particular jobs. Finally, PAMELA is primarily a collection of problem situations not of models for describing them; this restriction reflects several factors.

1 There are no 'right answers' to real problems.
2 Good models take a lot of space and time and merit fuller publication.
3 Good problems are harder to find than reasonable models.
4 If students are to tackle them 'afresh', the existence of standard solutions is inappropriate.

This having been said, there are suggested solutions and references for many of the problems.

PAMELA is specified at four levels:

1 title
2 essence
3 specifications
4 analyses.

Title

This sector contains a short title which identifies the problem (e.g. travel to school) and catalogues information on:

1 Subject area, e.g. economics—money, time.
2 Minimum ages, for 10th percentile and for average ability, at which it is profitable to tackle the problem.
3 Interest level, see page 8.
4 Time scale in hours.
5 Specifications—number, and number of questions in largest.
6 Analyses—number, number of models in largest.
7 References (not complete).

Essence

This sector contains a short description of the problem situation, for example,

'Compare various ways in which you might travel to school.'

Specifications

These are a sequence of 'well-defined' questions that may help to lead the student through the problem, identifying various facets of it. A given problem may have several specifications, though we keep only those that seem to have a significantly different approach. Where associated analyses are known to exist in published form, references to them are included.

Analyses

These are detailed discussions of the problem, including the formulation, solution, interpretation and validation of a sequence of models of various aspects of the problem. They will, we hope, normally be published in books or appropriate journals such as the *Journal of Mathematical Modelling for Teachers*. When an unpublished analysis exists in PAMELA this is noted in the specification.

Subject areas used by PAMELA

The subject area classification includes the following:

Economic

Money – personal	– wages, expenditure
– national	– GNP, inflation, exchange rates
Time – personal	– work and play scheduling
– at work	– efficiency
Resources planning	– travel to school, holidays

Physical

Mechanical	– traffic, sport, machines
Energy	– fuel resources, heating
Pollution	– litter, traffic, crowds
Popular science	– space, weather, disaster

Human

Domestic	– chores, possessions, house
Two-person interactions	– friends, relations, sex
Large group behaviour	– crowd reaction, control
Interactions at work	– friends, bullies, authority
Learning	– style, efficiency
Leisure, arts, sport	– music, colour, tactics, design

Biological

Physiological	– food, sport
Medical	– epidemics, innoculation
Reproductive	– genetics, contraception

Access to PAMELA is freely available to those involved in the development and teaching of mathematics and its applications. Listings from the collection of problems will be sent on request; equally contributions to the collection are solicited. The most important element of the contribution is the Essence—the brief statement of the problem situation; Specifications and Analyses are *not required*, but represent a separate contribution. The building of the collection depends on the range and ingenuity of those who contribute to it. Contributions and requests for listings should be sent to PAMELA, The Shell Centre for Mathematical Education, University of Nottingham, Nottingham NG7 2RD, England. All contributions will be acknowledged.

3.2 Teaching and assessment

Applications of mathematics

It will be clear to the reader who has reached this point that in facing his pupils with real problem situations, more is demanded of the mathematics teacher than in his traditional role. He may feel that this is simply not his job. We suggest that it should be, for two different reasons—not only is it important that pupils should acquire the ability to use their mathematical skills, but also that the pay-off in better understanding of the mathematics involved makes the effort worthwhile. In Section 1.3 we suggested a specific teaching approach; in this chapter we shall talk about various modes of classroom organization which may be used, and about the problems of assessment and some solutions.

Although the investigation of new situations is an important element in realistic applied mathematics, a substantial part of the time devoted to applications will always be spent on teaching standard models of standard situations. These can be, and probably should be, largely taught and assessed in a traditional way, ensuring that pupils can apply the models successfully in minor variations of the situation that they have been given in class. We would only advocate two changes of emphasis:

1 the choice of problem situations that are as interesting to the children as possible

2 careful discussion of the assumptions of any model with alternative possibilities and the reasons for the choice of a particular formulation.

So for example, we would use cars or people in studying kinematics, while a discussion of simple and compound interest is better done within the context of borrowing from a hire purchase company or a bank rather than stocks or shares. We have tried in this book to provide a nucleus of ideas of this kind but the class itself will provide more for the observant teacher to seize on (see page 16).

The introduction of *new* problem situations is more demanding of both pupils and teacher. Instead of just following the teacher's explanation, the class is being asked to go out on its own; the teacher, for his part, must be prepared to follow wherever they may lead, providing encouragement, suggestions and, on occasion, rescue from a predicament not of his own choosing. Given the difficulties that many children have with normal mathematics, how can we hope for success in realistic problem solving?

We have stressed that the problem situations which pupils are to tackle on their own must be much simpler than could be taught to them at the same stage. Secondly, by choosing familiar situations and tackling them in an informal way we bring into play other experience and understanding—so-called commonsense—which is so often left outside the mathematics classroom. Indeed, our objective might be phrased as: *the building of mathematical skills into commonsense.* Thirdly, since there are no 'right answers' but only more or less understanding, some initial progress can usually be made quite easily, verified and if necessary corrected by observation or commonsense, giving encouragement to push on further. For the teacher too, the situation is not as forbidding as it seems, though some changes from the traditional approach are essential. The clean certainty of much pure mathematics, which is part of what attracts people into mathematics teaching, is not present in the relation between the mathematical model and the real world—this will usually be neither right nor wrong, only more or less useful. The good applied mathematician must know which of the steps in his reasoning are purely mathematical, and therefore secure (in principle!), but much of his time and attention will be devoted to the other aspects of problem solving. It follows that the teacher:

- should be prepared not to know and to be wrong; (This departure from the traditional role of the mathematics teacher requires a good deal of self-confidence, particularly to begin with, because one must not be wrong or stuck for too large a proportion of the time. However, the teacher will usually be able to cope with any line of attack that is practicable for the pupils— when in difficulty the problem should be passed back to the group with some suggestions for a new line.)
- should be prepared to build models by intelligent guess work, and to follow pupils' models that may be of dubious validity, gently guiding improvement;
- should lean on observation and commonsense to decide what is successful and what is not in formulating and validating models.

All this demands ingenuity and above all, flexibility. Most people who try it find it enjoyable and satisfying, after an initial nervousness that fades with experience and achievement. This applies equally to pupils and teachers, though the former are often quicker to respond! To spend about 20 per cent of mathematics time in this sort of activity is a valuable leavening; its spirit is liable to spread wider as a healthy attitude of mind.

The detailed organization of real problem solving is still very much a matter of experiment and, as in most schemes, different teachers will find different ways to suit them best. We have already described the methods of the USMES project which seems to provide an extreme in each of the relevant variables—broad challenges tackled by the whole class over a long time scale at primary level, the opposite pole being the individual child working on a well-posed problem and interacting only with the teacher. Our suggestions in Sections 1.3 and 3.1 lie somewhere in between. We now discuss in more detail some of the most important variables.

Group size

Although committees never create anything, there is no doubt that the process of generating ideas is helped by discussion, and the size of the modelling group is a critical variable in teaching. USMES, as we have seen, starts with the whole class tackling a particular 'challenge'; then they divide into smaller groups to work on separate aspects. Regular reporting to the whole class leaves the broadest possible avenue for ideas and comments—creation and criticism. Conversely, the demand on the individual child is diluted if the group is too large and one worries about 'passengers'—USMES responds that each child contributes what he has to contribute, but teachers may be sceptical, particularly in the British classroom where conformist social pressures are perhaps less effective. Smaller groups of four to six have been found to work well, but there is a lot of room for experiment both in the size of the group and in the balance of ability and temperament within it. (It should not be assumed that the ability to solve real problems is simply correlated with conventional mathematical ability.) In some circumstances the 'buzz' type of approach, in which the children begin by considering the problems themselves and gradually build up co-operative groups through discussion, may be valuable.

It is fairly clear that different parts of the problem-solving process ideally require different-sized groups. While ideas generation can well be done in a large group, the detailed working through of mathematical arguments is probably best done individually for later presentation to, and checking by, the group. Between these extremes, detailed planning may well be done in pairs or trios. These factors, which need further investigation, raise problems of organization which will lead to compromises for the convenience of the teacher.

Organization and hierarchy

Much of the challenge of real problem solving lies in imposing order on a fairly chaotic collection of ideas, impressions, measurements and individuals. The organization of the group is the framework for the organization of information. As we have said in making suggestions earlier, no definitive scheme exists, but certain features seem to be important. Record keeping is essential, so a 'recorder' is needed in each group. From the wild ideas of the first brain-storming session to the description of the most fully-developed model and its predictions, the essentials must be written down, with sifting at each stage so that the information is available. Because experiment is so expensive in time and material, its planning and results should, particularly, be carefully considered and recorded. Though the structure of such groups is essentially egalitarian, a 'chairman' responsible for keeping the discussion flowing smoothly is helpful, at least in the early stages. The teacher can fulfil either or both of these roles, allowing him to control the progress quite delicately by selecting from the ideas presented; this may be useful to build early confidence but should not be regarded as a satisfactory general method of working; ideally the teacher is a mixture of interested spectator and outside consultant.

The teacher's role

We have already made some general comments on the role of the teacher. The principle, as so often, is to leave as much to the children as possible but to inject enough to prevent discouragement. It is particularly important to remember that the pupils are exploring new ground so both the rate of progress and the final achievement will be much less than when following a teacher down a well-established path.

There are two schools of thought on the preparation of the teacher for a particular problem. One is that the teacher must approach the problem as fresh and unprepared as the pupil or a false situation ('He knows and won't tell us.') is created. The price of such monasticism seems to us too high—it precludes the re-using of a successful problem in a field where good problems, rich with possibilities, are not easy to find. It underrates the professionalism of the good teacher in establishing a working relationship with his class. We have found with groups of a range of ages that a detailed knowledge of all facets of a problem allows guiding remarks to be made with a much more controlled effect. While each such injection is in a certain sense a tactical defeat, one cheerfully accepts that they will occur repeatedly in the slow process of building up the problem-solving skills and confidence of the pupils. On the other hand, it is both valuable and stimulating for an experienced teacher to face a completely new problem situation in the classroom from time to time. Often a fruitful area for this may arise from class discussion of other problems.

The teacher may need to help a group in any of the stages of problem-solving described in Section 1.2 and, in particular, to change gear from one stage to another—we all tend to enjoy a well-defined successful activity and to persist with it beyond its point of essential usefulness. Of all the phases of the problem-solving process it is perhaps the distilling of the first substantial mathematical statement from the collection of variables and ideas that is the critical point—plenty of time should be given after the need for this is pointed out, but a rescue may be necessary. Choose to inject a very simple model statement—it can be built on later.

When the children are stuck, two tactics are useful:

1 Lower the level of abstraction—if they can't make a general argument, suggest they try a specific case; if algebra is too hard, use arithmetic or a graph or build a physical model.
2 Get more information about the situation—even if the model doesn't help them understand what's going on, it will probably suggest information that might help.

The length of the project

The length of time to be devoted to a particular problem will vary with the problem situations, but will always be at least an order of magnitude longer than the few minutes that are required for a repetitive technical exercise. At the other extreme, large projects like the USMES challenges may take many weeks at several sessions a week. The initial consideration and discussion, where no real progress is apparent, needs a lot of time, often taking as long as the completion of a satisfactory initial analysis.

On the whole it is a good idea to devote more than one class period to a problem because the mind goes on digesting the ideas between sessions, which should not, however, be too far apart or things will be forgotten. Two periods on successive days at fortnightly intervals, with some private work, experiment or writing-up in between, is a reasonable rhythm to try—but the double period for the first session can be an asset later when the reports on earlier problems may provoke discussion and further ideas. We need to know more about how these various patterns work.

The roles of video

Interesting possibilities exist for using film and other visual material in teaching of this kind.

1 Film can be used as a way of posing really new problems in a relatively unpredigested, unstructured form. Language contains so little information that any description of reasonable length must be highly structured, picking out the essential features in a way which often prejudices the crucial initial phase of problem solving. We have avoided this mainly by using very familiar situations which have only to be identified, not described. However, if you want to learn how to tackle somewhat less familiar problems there can be an advantage in some cases in presenting them visually. A short film of a high jumper, a pole vaulter and a long jumper in action contains a lot of the information needed to understand the similarities and differences in their sports, but without prejudicing the analysis. This would be impossible to do in words (although we could move somewhat beyond standard examination questions by including a lot of redundant information, for example).

2 Film can provide a useful time-saving substitute for experiment in validating and improving models, particularly if single frames can be shown and measured. Traffic problems are an obvious example.

3 The use of video recordings of problem-solving sessions is a simple and direct way of showing how they may go, but which is useful in the training both of pupils and teachers. USMES has used live video to connect two classes working on the same real situation.

The development of such resources requires facilities beyond those available in most schools. It should be pursued more centrally.

Assessment of modelling skills

We shall not say very much in this book about assessment. This is not because we underrate its importance in mathematical education, but because techniques for assessing modelling skills are not well developed.

Let us first discuss those which arise naturally from the problem-solving process, particularly the 'final report'. The writing of such a report by each pupil, describing the problem, explaining the approaches considered, the reasons for choosing those that were explored in detail, the results of their detailed study and the evaluation of their usefulness, is a valuable but demanding activity. Apart from the mathematics involved, it requires skill in the use of language and the martialling of argument that are well worth development. Standards will tend to be low in both areas but both aspects should receive attention and credit. Priority should probably be given to comprehensibility of argument rather than to its literate presentation.

Such reports can test the pupils' understanding of all the aspects of the solution that we have discussed though, if there has been group activity, it does not show the individual's contribution to solving the problem, particularly the more creative aspects of model formulation and solution. We do not think this limitation should be regarded as serious—the range of skills being assessed is already much wider than in most mathematics examinations where originality is rarely demanded. A more serious difficulty is that lack of skill in one area, say the use of language, may prevent a pupil showing his understanding of the problem; this may be partly overcome by discussing the report with the pupil, which incidentally exposes undigested plagiarism. A full written report by an individual pupil should only be demanded occasionally because of the time involved, but a comprehensible account of the problem and its solution is an essential element in modelling; for this and other reasons there are advantages in using group reports much of the time.

Other ways of assessing pupil progress are also useful. If the teacher is involved in the initial discussion of the problem, he will have a pretty clear idea as to who has contributed what and this can be used in assessment. This approach is not 'fair' in that speed of thought and outspokenness are favoured, but similar biases are present in all assessment systems and these particular ones may be closer to those the pupil will face in real life. Further contact with the group as the problem proceeds will contribute to building the teacher's informal impression. The more straightforward activities of calculation or data collection can be assessed directly if desired.

The direct testing of particular skills in model formulation, interpretation and validation is possible, but research is needed to ensure that these are indeed crucial components in being a good, realistic applied mathematician. The ability to generate ideas, variables or relations when faced with a problem can be tested separately—the number and range of ideas are clearly separate factors. The ability to choose helpful and sensible lines of attack, to carry them through, to interpret answers, and to divide searching tests of a model's validity are all under study. As practical possibilities for assessment, these are for the future—for the moment it seems sensible to de-emphasize assessment and confine it mainly to overall success with the problem. Certainly it will be most unfortunate if assessment intrudes on the problem-solving process itself, particularly in its more creative aspects, as this could easily damage the supportive role of the teacher and distract the student from the main purpose—understanding the situation. For the moment, therefore, we simply wish to give some rules of thumb:

– Regard assessment mainly as feedback on the success of the learning process for each pupil.
– Concentrate mainly on the individual's ability to organize and present a useful, coherent (preferably written) argument, using mathematics when it helps (*not* on who contributed what to the group development of the argument).
– Don't expect too much or take the results of the assessment too seriously.
– In discussing the progress made, the central criterion is the usefulness of the understanding gained.
– Don't encourage the children to drag in more mathematics at any stage than has a probable pay-off in their hands.

3.3 Learning more about teaching modelling[*]

Now let us look in a little more detail at the way we go about the process of tackling a real problem situation. We outlined our model of modelling in Figure 1.1, giving a more detailed picture in Figure 1.4, and discussed the four major aspects (formulation, solution, interpretation and validation), and the simplification and improvement phases of the looping structure. Each of these is a pretty complex activity worth

---

[*] This section was written with Vern Treilibs.

further analysis. The mathematical solution process has been the subject of a fair amount of research in mathematical education; we shall not discuss it further here. We shall say something about interpretation in Chapter 4 when we discuss the translation skills which it involves. Here we shall concentrate on formulation which is perhaps the central activity in modelling, though we shall also add a few words on validation. We look first at the complex synthesis of modelling activities that constitutes the formulation process.

The formation of mathematical models of new situations has not until recently been taught, or expected of students, except at a research level where the unstructured apprenticeship approach to teaching (sitting next to Nellie) and assessment by thesis have generally been used. Figure 3.2 describes at a deeper level a model of the 'formulate' box of Figure 1.4. It must be said that in both cases the model of modelling that we put forward is based largely on introspection, and research on its validity is still in the early stages. We distinguish two different types of model formulation activity, which we call *analytic* and *descriptive* modelling. The latter is concerned with collecting, organizing and assimilating experimental information about the situation, whilst the former concentrates on developing an understanding of the underlying structure of the situation being modelled. For example, in modelling the layout of goods for sale in a supermarket, a statistical survey of existing supermarkets would constitute a descriptive modelling exercise, whilst the finding of the relationships between, say, position and selling rate

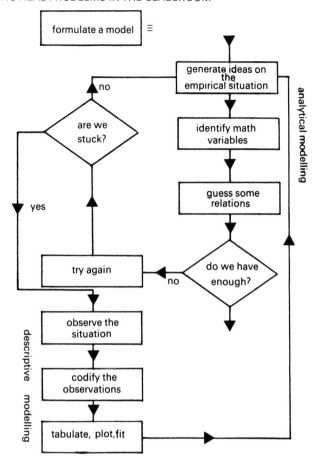

FIGURE 3.2

would be analytical modelling. Although there are those who believe that
the descriptive process is the essential activity in model building we, with
Karl Popper, have no doubt that in thinking about a new problem we
always start from our prejudices, which are rough analytical models; we
regard this as self-evident, since the decision on what observations to
make from the myriad possibilities needs a model to guide it.

So we begin with *analytic modelling.* There are at least three types of
activity that are important:
- generation of ideas
- selection of ideas
- development of ideas
where the ideas may be:
- areas of interest in the situation
- variables
- relations between variables.

Added to these is the identification of the specific questions to be answered in the often ill-defined problem area being considered. This activity is sometimes overlooked by problem solvers, usually with disastrous results. The relative importance and difficulty of these activities is not well understood, but they can be looked at separately and this is being done in various ways. It seems likely that the generation of useful relations is a demanding activity. In the generation of aspects/variables not only their number but the range of different types of possibilities explored is obviously important. We believe this analytic modelling loop remains active until a model rich enough to predict something is found (when we proceed to the solution box in Figure 1.4) or until we are unable to find such a model. (This may also occur at some later stage of the improvement phase.)

At this point we may resort to *descriptive modelling*. We have by now found variables that we regard as important, with perhaps some relations between them, but we want more. We may determine such relations empirically by observing the system and finding out how the variables depend on each other. These empirical relations may suggest underlying analytic structures. For example, we may have no idea how the positioning of different goods in a supermarket is decided; if we find that the most commonly bought products are farthest from the door, this may suggest some possibilities—that the choice of position aims to maximize purchases by the customer (by making him walk past as many goods as possible) and is not for his convenience. Equally we may simply lack factual information with which to launch our model—the rate of purchase of particular products, or the interest rate on bank loans, or the average speed of a bus. We could proceed to model algebraically with the unknown quantity as a variable, or we could find its value by 'observation' to keep the analysis simpler and more specific—this latter method is descriptive modelling.

Having formulated the mathematical model, appropriate mathematical techniques can be applied to it and the solutions interpreted. These processes are usually fairly straightforward, but the final phase, validation, requires the generation of ideas for the *testing* of the model, and in some situations validation can make substantial demands on creative ingenuity.

We conclude this section with some comments on the situation in the 'modelling classroom'.

1 Modelling takes rather different forms according, in increasing order of difficulty, to whether it is built:
 – directly on given numbers
 – on numbers found experimentally, or looked up
 – on numbers known only after an analytic model is built
 – on ideas rather than numbers.

This ordering is related to our earlier advice to lower the level of abstraction when in difficulty. There is some evidence that students choose to work at the lowest level permitted.

2 It is an interesting and open question as to how far conscious awareness of the modelling process improves performance at it; the traditional plight of the centipede who over-intellectualized about the mechanics of his locomotion is a useful warning. It is not, of course, necessary for teacher or pupil to have a detailed picture of the formulation process, but it may help in building modelling skill by suggesting lines of progress when in difficulty. Our impression is that the rough and ready analyses we have described and the heuristic rules we have derived from them are helpful in teaching pupils to tackle real problems.

3 A comment about *externalization*—the revealing of one's thought processes. This is a crucial aspect of teaching and learning since the pupil is partly seeking to imitate the teacher's thinking, while the teacher is trying to follow the pupil's thoughts in order to diagnose his confusions and help him progress. Neither is very well served by traditional teaching in this regard. The teacher in his prepared explanation produces a polished, nearly linear (language demands linearity, though pictures allow more variety) description of the material being taught which usually bears little relation to the complex system of inference from several viewpoints by which we seem to learn. The pupil, working silent and alone, reveals his thoughts on paper in an equally abbreviated form. Group discussion of problems in their early unstructured phase causes everybody to reveal much more; the danger is that the picture is initially confusing, but the sorting out of such confusion is the essence both of modelling and learning. Thus there is hope of rather special support to the ordinary mathematical learning process from these activities—as we have noted, the USMES evaluation studies on progress in basic mathematical skills gives results in accord with this hope. There is much to be gained from developing skills in group work in students; conversely the use of *ad hoc* groups can prove disappointingly inefficient.

4 Teachers adding a modelling component to their courses might bear in mind that current syllabuses are reasonably successful at developing the *critical* aspects of the modelling process, but rather poor at developing the *creative* ones. Consequently the main emphasis will need to be placed on developing formulation skills.

5 It is worth noting that no amount of modelling skill will compensate for a poor knowledge of standard models—and conversely!

3.4   Comment

There is, of course, no reason why this activity should be classified as mathematics, and collaboration with a like-minded numerate teacher in another subject may help by providing both moral support and class time. However, the same willingness to start from problems that are real to the pupils, to call on the whole of one's experience in tackling them and to fail quite frequently, are needed; other school subjects do not necessarily produce this breadth of outlook any more widely than in mathematics.

Some may be worried that what we have been discussing is another 'new maths', a wave of enthusiastic reform which will turn into an uncomfortable, alien orthodoxy which teachers will feel they must follow. This does not seem to us a real danger. Most of mathematics will continue to be taught as before, though perhaps with more emphasis on the practical and the useful than on the formal learning of mathematical structures which are not used by pupils at that stage. The teacher can decide the amount of modelling that he wishes to inject. The extra demands it makes call upon talent which we all possess—commonsense and some knowledge of the world; they do not involve the learning of more formal mathematics.

By now you, the reader, will have gathered that the tackling of real problem situations in mathematics lessons is still something of a pioneering enterprise, with a significant but still quite limited amount of experience to lean on. You will know by now whether you are tempted to have a go—indeed those of you who have stayed with us so long are probably mostly of this mind. If you do so, we also would like to hear of your experiences, good and bad, amusing and frustrating, so that we all may benefit. Detailed curriculum material designed to help teachers introduce some modelling into their current programmes is being developed at, and is available from, The Shell Centre for Mathematical Education at the University of Nottingham.

4

MATHEMATICAL
TECHNIQUE
FOR APPLICATIONS

4.1 Technique in applied mathematics

There is a fair amount of confusion in the use of the phrase 'applied mathematics'. It is used by us and many others to describe the use of mathematics in understanding phenomena, as in applications to mechanical, economic or everyday problems. But it is also often used to describe the mathematical methods which are valuable in such applications—for example, statistics provides a set of mathematical models useful in almost every area of application, and trigonometry is a set of mathematical techniques with applications in surveying, navigation, mechanical and civil engineering, and physics. It is worth seeing clearly the distinction between the application, which begins and ends with a situation outside mathematics, and the mathematical methods or techniques which are used in understanding it. We will spend some time talking about methods in a book on applications because the applied mathematician needs as wide a range of the former as he can command. Mathematical methods form the tool-kit which he uses in building models from the materials of the problem in hand; the more comprehensive his kit the more powerful he will be, provided of course that he can use the tools he owns! We have already emphasized that, in order to use a technique in applications, particularly in model formulation, it needs to be thoroughly familiar.

This book is not the place to discuss in detail the spectrum of mathematical techniques that may be covered in the 11–16 age range. (The main aspects of the current curriculum have recently been critically reviewed in *The Mathematics Curriculum* series published by Blackie for the Schools Council.) It is clear, for instance, that arithmetic, algebra and geometry form the basis on which applied mathematics rests; here we repeat only our central point—that a lot of explicit help and practice are required if a student is to build this mathematics into the equipment he uses in facing the problems of the everyday world. There is, however, a range of techniques not traditionally covered which are of great importance in applications—flowcharts and other graphical methods, the use of calculators and computers, simulation and optimization methods, are principal new areas which we shall discuss here, if only briefly. Equally, some traditionally useful topics are no longer taught in such detail—trigonometry is an example. The whole area of phenomena where there is a significant amount of random or uncontrolled variation, requiring the methods of statistics, is especially important and difficult.

Not everything can be taught. The range of techniques that each of us has will always be limited, but it is important that we know what we have at our disposal, and can use it. For the more effective and organized mathematician it is also valuable to know something about other methods and what they can do, so he can learn them when he needs them.

There is a difference of view about where techniques should be taught. They may be introduced in the context of a particular application, or taught in their own right with illustrations from various areas of application. The advantage of the first approach is that the motivation and pay-off are immediate and clear. Vectors taught in the context of balanced systems of static forces can be an instant success in both calculation and understanding. However, the separate teaching of methods with a variety of illustrations has a number of advantages. It makes it clear that the method is a general mathematical tool with a range of areas of application, and usually with links to other areas of mathematics. It also avoids confusion between the mathematics itself and the models of a particular situation, and emphasizes the varied applicability of mathematical methods.

In discussing the value of mathematical technique, it remains to stress the importance of reliability. In solving problems for which the answers matter, and which may form the basis of decisions affecting the life of the student, it is essential that, after checking, the accepted result is reliable in the sense of being free from blunders. This implies standards of precision quite different from the normal British examination pass mark of 40–50 per cent on a choice of questions, which guarantees no specific skill. In discussing this difficult problem of curriculum design, it may be useful to distinguish, say,

acquaintance	40%	'Have seen'
competence	60%	'Know how'
mastery	90%	'Get right'

The marks and descriptions are intended to be suggestive rather than a precise definition. On the whole, education in Britain at all levels has been content with competence, and for many students acquaintance has been a considerable achievement.

However, most adults have zones of mastery on which they rely quite heavily, as well as a wide area of acquaintance which provides richness, intellectual breadth and a knowledge of what could be learnt if it were needed. It is an important educational question as to whether, for each student at each stage, we should aim to specify a small area of mastery within the whole curriculum and to teach, test and refresh it at that level. This is probably the essential question at the heart of the debate on the core curriculum and basic numeracy, but it applies equally elsewhere, for example in the sixth form and at undergraduate level. Since we have little experience of teaching for mastery in this country, experiments will be needed to discover what can be done before deciding what should be done. Because mastery is not easy to achieve, it is vital that the targets be realistic—they will probably seem very limited and public acceptance of such targets may not be easy to achieve.

Fluency, standard models and creativity

In the context of applied mathematics, mastery means much more than a fluent accuracy in the mathematical operations themselves; it also requires a familiarity with the properties of the mathematical structures and operations in use, and with a range of the situations in the outside world that they can describe. It is no use being able to do a long multiplication if you do not know whether to add or to multiply. The well-known difficulty in teaching proportional reasoning for many children becomes less strange when one considers the following problem.

'If it takes 45 musicians 30 minutes to play Beethoven's 5th Symphony how long will it take 80 musicians to play Beethoven's 9th?'

The proportionality of two quantities is a potential mathematical model,

$$y = mx$$

if you like to use algebra, which may or may not be valid. All of us possess an array of such models and know a range of situations where each of them is useful; according to our sophistication we may be confined to models using only arithmetic and measurement, say, or we may have access to the immensely wider range of possibilities that the more general abstract forms of algebra, calculus and their developments make possible. It is known to be much easier to scale up or down by a factor of 2, or a small whole number, than to do general proportion problems; similarly a choice of 2 alternatives is much less demanding than a higher number. Whatever our range may be the models in it will only be useful to us if we know them and their properties well. The *Mathematics Applicable* course (see page 119) aims specifically to build up this fluency.

We should also consider further the role of creativity in learning mathematics. In the traditional mathematics curriculum the student is expected to show very little independence in his thinking; the problem has correct answers and correct methods of solution which have recently been taught. In this book however, we have cast the pupil in a much more independent role in choosing both the aspects of the problem situation to be studied and the means to be used in that study. We have talked in terms of a mathematical tool-kit from which the student selects the tools that seem best suited to the job in hand, and in most of our examples we have used only quite simple mathematical ideas and methods. The objective has always been the better understanding of the situation. However, in this part of the book we are concerned with mathematics itself. Can this same approach give the student an expanded role in helping to create and enlarge his own mathematical tool-kit?

We think it can and indeed that it will, because the exploration that is the essence of modelling new situations is also the basis of mathematical creativity. The investigation of particular cases suggests generalization; for the abler student such conjectures demand firmer verification, and perhaps sometimes even formal proof. This is of course the basis of the 'investigation' approach to studying mathematics itself, but in modelling well-understood real situations the student's mathematical intuition is firmly guided by the concrete embodiment which is being modelled. In this context reality is a crutch that can sometimes be abandoned and the infinitely wider range of fascinating possibilities suggested by the mathematics can be partly pursued. The situations in the *Mathematics Applicable* material exploit this approach, but on the whole they do so by leading the student. However, experience of such 'projective' modelling, in which one explores a wide realm of conceivable situations rather than real ones, is valuable training for any applied mathematician as well as being intellectually stimulating.

Translation skills

There is one set of mathematical skills, often not explicity identified, which are of such importance in applications that they merit special discussion—we call them translation skills. Mathematical information may be represented in many forms—descriptions in words and pictures, graphs of various kinds, tables of numbers, algebraic expressions—and each of these has its advantages. The ability to translate the information fluently from one form to another is therefore of great importance. The table shows what each of these skills is generally called. The arrows show intermediate steps which are usually used in some of these processes—for example in curve sketching it is usual to calculate and plot some points.

To \ From	Descriptions	Tables of numbers	Graphs	Algebraic expressions
Descriptions	—	measuring	sketching	descriptive modelling
Tables of numbers	reading	—	plotting	fitting (descriptive modelling)
Graphs	interpretation of graphs	reading-off	—	curve fitting
Algebraic expressions	formula recognition	tabulating or computing	curve sketching	—

While some of these skills, such as the plotting of graphs, occupy a prominent place in the mathematics curriculum, there are others that are equally important in practice, which have been neglected. Almost all of these involve modelling skill and are by no means as easy as one might think. We shall illustrate this by discussing one of these processes, the interpretation of cartesian graphs, in some detail. We choose it because of its importance in understanding information presented graphically in newspapers and other aspects of everyday life, and because we have recently studied it in association with Claude Janvier; we would expect many of the same features to apply to other processes such as the interpretation of train timetables and other tables of numbers.

Figure 4.1 shows the speed of a racing car as a function of the

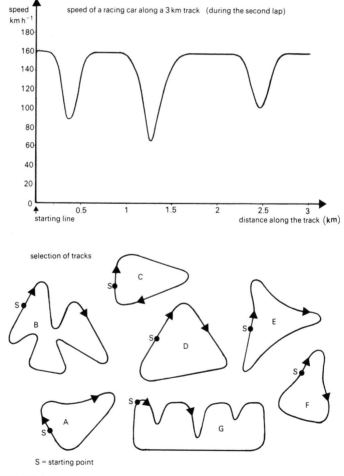

FIGURE 4.1

distance along the track. Many pupils find it hard to describe what the graph represents—that each dip in the speed corresponds to a bend in the track, with the left-hand side of the dip representing 'slow-down' and the right-hand side corresponding to 'speeding-up', that the deeper dip in the middle represents a sharper bend so that the starting line is between two gentler bends, and so on. Faced with the seven alternative possible shapes of the track, many initially choose track *G*, and all find the selection process difficult. These pupils could successfully plot points on a graph, and read points from it, but had no experience in interpreting *global* features such as the level, slope, dips and peaks, and relative positions of different parts of the curve.

This is not surprising. The cartesian graph is a subtle and ingenious convention—one must understand first that a function maps every point in the domain R_1, say, into a point in the range R'_1, see Figure 4.2, and that for each point x in R_1 you can draw an arrow leading to the

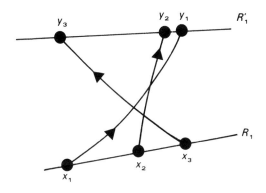

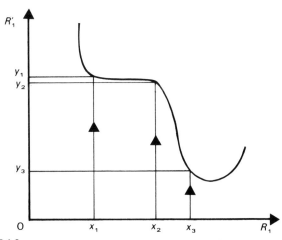

FIGURE 4.2

corresponding point y in R'_1. The cartesian graph then notices that if you choose always to draw your lines representing R_1 and R'_1 at right angles with equal spacing along the axes, and that, further, if you make all the arrows consist of a vertical line through x and a horizontal line through y, each arrow can be specified simply by marking the corner where the vertical and horizontal segments meet. Then all the infinity of possible arrows are given simply by the line passing through all the corresponding corners together with the recipe:

> 'From any point x you go vertically until you hit the curve, then horizontally to the y axis to find y.'

—a masterpiece of compression and economy.

This analysis still relates only to points. Fluency in the interpretation of global features of graphs will only come with explicit teaching and practice. There is not space here to describe the methods that are being developed for doing this, but once the need is recognized we believe that teachers will be able to develop their own.

Flowcharts, graphs and so on

A number of mathematical techniques which have been introduced fairly widely in the last 15 years are powerful in application, and seem to be accessible enough to many children to be useful. We discuss some of them in detail later but others are worth mentioning.

Flowcharts have been referred to already. They seem to have been absorbed into the mathematics curriculum painlessly and profitably as a way of describing algorithmic procedures. How far they are useful for applied mathematicians in *devising* algorithms, as opposed to describing them, is one aspect of the process of problem solving that is still controversial (it is sometimes called structured programming). However, they are clearly a useful and accessible tool.

Tree diagrams, and more general graphs, also have a firmly established place. The book *Counting and Configurations* in *The Mathematics Curriculum* series makes clear their value in discussing combinatorial problems, including those from probability. The graphs of processes in the 'critical path' method of scheduling work on projects (see page 128) are only one of a number of applications in this book. Because pictures have a greater impact on most of us than numbers or algebraic expressions, graphs of all sorts can play a key role in making the structure of problems clear, though as we have said, skill in interpretation must be learnt.

Other recent introductions to the curriculum do not have such a clear pay-off in practical terms. Matrices are an extremely powerful tool in applied mathematics of many kinds but their pay-off is only apparent for systems of some complexity, involving several variables. These are beyond the scope of most O-Level curricula and the 2×2 cases treated there aim on the whole to provide insight rather than extra power—the practical problems can usually be tackled equally well by graphical or elementary means. For this reason, the premature introduction of such topics may be a disservice to mathematics.

Mathematics Applicable

While all mathematics courses make some implicit use of modelling skill and intuition by reinforcing mathematical ideas with illustrations from outside mathematics, the Schools Council Sixth Form Mathematics Project has developed a new approach in which modelling plays a central role. This project, directed by Christopher Ormell, produced a series of books for a one- or two-year sixth form course which is published by Heinemann under the general title of *Mathematics Applicable*; an A/O-Level examination for this course is offered through the University of London Examinations Board. Each book in the series tackles a particular topic area—the aim is to build up understanding and skill with the mathematical techniques concerned by using the mathematics to model a wide range of practical situations; these are often of a whimsical rather than a fully realistic kind, which for these purposes is a good thing, since it allows a profusion of examples. However, the course and examination also aim to teach the student to use his acquired understanding of the mathematics to model real problems of interest, though these are mainly presented in the form of well-posed questions.

As a training in mathematics from an applications point of view this approach seems to offer real advantages; it will be interesting to watch its progress and see if, and how, its influence spreads into CSE and O-Level examinations. This has already begun—those interested can learn more from the Mathematics Applicable Group at the Department of Education, Reading University. Here we quote from the introduction to the Teachers' Guide:

The Series has been written mainly with non-specialist students aged 15–19 in mind. *Mathematics Applicable* offers a way in which they can be helped to keep their mathematics going during their last two years at school or sixth-form college. It offers a solution to the dilemma illustrated notionally in Figure 1.0 and 1.1 opposite. As students grow older and begin to enter the more sophisticated levels of thinking of their major subjects, their need to understand mathematics tends to increase. This is shown notionally in Figure 1.0. Yet, using conventional formal material and adopting standard approaches to the subject, their attitudes towards mathematics often tend to fall off. This is illustrated notionally in Figure 1.1. The two figures, taken together, constitute a dilemma: how

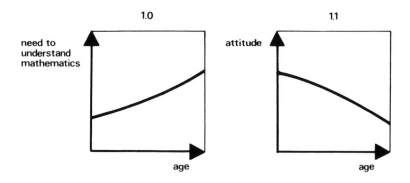

to present mathematics to the non-specialist student in the last two years at school or sixth-form college?

Mathematics Applicable tackles the dilemma by adopting a new approach. It aims to make mathematics 'relevant', and hence interesting, for the student. It does this, firstly by pitching it entirely in a modelling context, and secondly by using the modelling to discuss consistently interesting problems. These problems are basically about the *possibilities* of the real world: the possibility that the Ancient Egyptians used mirrors in their tomb painting, the possibility that an open-air diving bell could be produced, the possibility that fishing limits will increase, or that flats will be built to an immense height, or that Survival Stations will be placed across the Sahara.

Most students find problems of this *projective* type motivating as individual problems. But in order to obtain the full value of the material it is necessary to consolidate carefully and to build on this initial interest. To do this a particular teaching approach is needed. It involves consciously distributing one's teaching emphasis in a subtly different way. In fact it involves adopting a new teaching stance.

We believe that this new kind of teaching stance will become increasingly relevant at all levels of school and college mathematics during the coming years.

The new emphases are these:
Interesting the student by discussing the kind of possibilities which have a dynamic of their own. Mostly these consist of the interesting *ifs* of ordinary life and technical development. Given the right topics of this kind, we can set the student thinking, and if one handles the teaching process skilfully, one can ensure that quite a lot of *mathematics* gets carried along in this flux of thought.

Connecting mathematics with the real world. We need to practise over and over again the basic form of the link between mathematical symbols and the real world. The skills involved here are called the 'translation skills': translating from situations to equations (abstraction) and from equations to situations (interpretation).

Consolidating the methods, concepts and results which the student learns. This is vital if students are to manage the abstraction and interpretation aspects of their modelling successfully. The syllabus around which the *Mathematics Applicable* Series has been built is deliberately modest, in order to give students the best possible chance to achieve a firm grip on what they learn. Consolidation produces confidence, and confidence is needed if students are to use their mathematics effectively to tackle new and previously unseen problems.

Although the course is perhaps primarily concerned with learning mathematics through modelling, the tackling of real problem situations is an important part of its philosophy. Until recently the assessment of the course included an essay on a project and these were normally mathematical discussions of practical situations; we regret that the Examination Board has discontinued this element, providing as it did encouragement for the teacher to tackle real problems in the course.

Nonetheless, the teachers' workshops held by the Mathematics Applicable Group show an active interest in this aspect. Here are some problems discussed in one such workshop.

1 What should be the spacing between passing places in country lanes? (Assume a combination of straights and curves, the length of the straights presumably being a factor in the answer.)
2 The rate of cooling of a pottery kiln/cooker oven (electrically fired)—at what stage can you switch off so that the residual heat will finish the process?
3 The relative advantages/disadvantages of packing camping gear on a roof rack or in the boot of your car.
4 The tilting of wooden chairs by students—repeated stresses cause ultimate collapse. What is the optimum life of a chair and the cost of replacement?
5 Roundabouts versus traffic lights.
6 The pros and cons of a 7 or 8 period school day.
7 The optimum size of a new newspaper.
8 Optimum driving speed for maximum traffic flow.
9 The rate of increase of candidates for maths A/O exams.
10 Smoothing out lumps in thixotropic paint.
11 Is 3-D knitting worth learning?
12 The rate of cooling of hard-boiled eggs after they leave the heat.
13 Rugby: injuries in scrummaging; depreciation of 'aggro' with time, due to player fatigue.
14 The frequency of recharging an electric razor.
15 How big should a spider make its web?
16 How many tills should Tesco have at a given time to cope with the customers?
17 The best serve in a tennis game.
18 Southampton has a rush hour traffic scheme which allows only a

certain number of vehicles into the city centre. Investigate the traffic flow, numbers of parking spaces needed, size of queueing areas outside city limits.

19 How often should I spray to control thrips and aphids as economically as possible, given that I want a maximum infection of not more than 20 pests per plant at any one time?

20 Optimum design of cream carton (and opening strip) for individual cups of coffee.

4.2   Optimization

A large proportion of the decisions in life are attempts to make the best of a situation; mathematics can often be valuable here, particularly if there are many possibilities or if there is a clearly defined measure of desirability. This branch of applied mathematics goes by the general name of optimization. It uses a wide variety of mathematical methods, depending on the nature of the problem—some of these are accessible to the 11–16 age range. More fundamentally, it is important that the *principles* of optimization be understood at a simple level so that the student can later appreciate what is involved in so many of the decisions that will affect his later life.

In every case we seek to achieve the largest values of some desirable quantity (e.g. maximum income or satisfaction), or the smallest of an undesirable one (minimum cost or pain), which is mathematically an equivalent problem amounting to a change of sign. The choice of what is to be optimized is a modelling problem—finding the mathematical entity f which corresponds to our real objective as a function of the variables of the situation, $\{x_1, x_2, \ldots x_N\}$. The variables may be truly independent or they may have constraints connecting them. Each constraint can be expressed in terms of some implicit function of the variables, $g_n(x_i)$, $h_m(x_i)$. There may be equality constraints, for example,

$$g_1(x_1, x_2, x_3, \ldots, x_N) = 0$$

$$g_2(x_1, x_2, x_3, \ldots, x_N) = 0 \qquad (4.1)$$

$$g_3(x_1, x_2, x_3, \ldots, x_N) = 0$$

which effectively reduce the number of variables by n, or inequality constraints which we may write as

$$h_m(x_i) > 0 \qquad (4.2)$$

which simply require the variables to lie within certain domains, for instance some of them may simply have to be positive or less than some upper limit. We shall see an example of an equality constraint in equation 4.4 and an inequality constraint in equation 4.8. We may find an optimum in one of three ways: by enumeration of possibilities, by finding stationary values or by endpoint optimization.

Enumeration of possibilities

If *all* the relevant variables are discrete and take on a limited set of values we can work out all possible values and find which is the best. For a few alternatives this is a good method but it can easily become tedious. It would be used for example to decide whether to use coal, oil, gas or electricity for central heating, or whether to seek a further educational qualification or not.

There are of course many other examples. In principle the enumeration method can be extended to continuous variables by 'discretizing' them, i.e. by choosing a set of typical values sufficiently close for the purpose in hand.

For instance, to find the largest rectangular rabbit run that can be surrounded by a total of $T = 12$ metres of wire fencing, we can try varying the width we assume for the run and calculating the area

$$A = WL \qquad (4.3)$$

remembering,

$$2W + 2L = T. \qquad (4.4)$$

Width (*m*)	0	1	2	3	4	5	6
Area (*m²*)	0	5	8	9	8	5	0

This suggests that a width of 3 metres, corresponding to a square run, is the best. An electronic calculator makes a fairly thorough exploration quite easy—in this case we can easily check that 2.9 and 3.1 metres do indeed give areas smaller than 3, and equal.

A rough graph of the points will produce an approximate continuous model, and a look at it suggests some interesting questions which we will return to later (Figure 4.3).

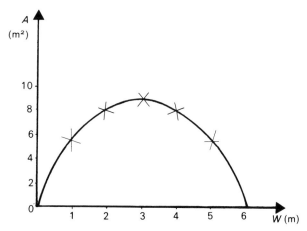

FIGURE 4.3

This approach can be used for any number N of variables, but it soon gets very tedious—10 points in each of N dimensions gives a total of 10^N evaluations to make and compare. The method of critical path analysis is another technique for optimization based on enumeration of possibilities. It is discussed on page 128.

Finding stationary values

If the variables and functions are all continuous and differentiable then we can use the property that the top of a smooth hill is flat (i.e. horizontal and level)—more precisely the gradient is uphill (positive) as you approach the top, and downhill (negative) as you leave it, so the gradient will be zero at the top. The finding of gradients of general algebraic functions requires calculus and so this method will only be acceptable to most 16-year-olds via a graphical approach, though many problems can be handled by completing the square or otherwise (where calculus is taught it is usual to do maxima and minima as an early illustration). The rabbit run problem now requires the maximum of

$$A = WL$$

where $2W + 2L = T = 12$ metres, so that

$$A = W\tfrac{1}{2}(12 - 2W) = W(6 - W). \qquad (4.5)$$

Now the gradient or slope is

$$\frac{dA}{dW} = 6 - 2W. \qquad (4.6)$$

This is zero when $W = L = 3$ metres. The diligent student will then check that this gives a maximum. He will probably not worry about the minimum area which method (1) clearly reveals—when the width is 0 or 6 metres. It clearly is not given by equation (4.6); we return to this point later.

The calculus of more than one variable is certainly not accessible at this age range, but the graphical method, which allows us to see at a glance the maximum of A in Figure 4.3 can be extended to a function of two continuous variables by representing it as a three-dimensional model, or as a contour map and again looking for the hill-top.

End-point optimization

Often the best, or worst, solution will not be a flat hill-top. In the rabbit-run problem, the minimum (zero) area arises because the width and length of the run have both to be positive—if we look at the graph in Figure 4.3 of $A(W) = W(12 - 2W)/2$ it continues down to minus infinity as $W \to \pm$ infinity, but as a model it only has any meaning if $0 < W < 6$. Such *inequality* constraints often produce optima, which we call end-point optima, when the constraint is reached.

Linear programming is a set of methods for use when both the function to be optimized and the constraints are linear functions of the variables. It is widely taught. We discuss it briefly here in order to relate it to other methods of optimization.

We illustrate what we mean with the simple problem of feeding the dog—we give it tins of Growl ('rich in protein') and packets of Crunch ('nutritious', i.e. lots of calories) and we want to keep the cost down. Crunch provides the cheapest way to give the dog the necessary calories C, but it needs P of protein too, so some Growl is indicated, but this costs more and also has calories in it. This chaos is all sorted out by choosing to give the dog x_1 of Crunch, which has p_1 of protein and c_1 of calories and costs m_1 per unit quantity, and similarly x_2 of Growl, which has p_2, c_2, m_2. We want to minimize the cost per week:

$$T = m_1 x_1 + m_2 x_2 \qquad (4.7)$$

Subject to requiring:

$$c_1 x_1 + c_2 x_2 \geqslant C \qquad \text{(enough calories)} \quad (4.8a)$$

$$p_1 x_1 + p_2 x_2 \geqslant P \qquad \text{(enough protein)} \quad (4.8b)$$

$$x_1 \geqslant 0, x_2 \geqslant 0 \qquad \text{(you can only feed it not milk it!).} \quad (4.8c)$$

With two variables, geometry is easier than algebra and Figure 4.4 shows the solution. The last two conditions say we must be in the top right-hand quadrant shown. The nutritional conditions are satisfied in the areas above and to the right of the equality lines C and P, so the area double-shaded is the one that does not starve the dog. (Strictly, the boundary lines are included in the permissible areas in the way we have written the condition, but this mathematical distinction is quite unimportant for the practical purposes of the situation since we are clearly incapable of measuring food with infinite precision, and even the definitions of c and p are approximate.)

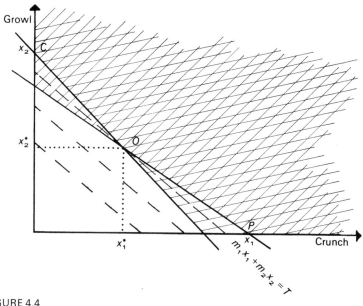

FIGURE 4.4

Now to find the minimum cost we notice that the lines of constant cost, shown dashed, each correspond to a fixed cost which increases as we move upwards and to the right. We therefore want to be on a line as near to the origin as possible but within the double shaded area. The point O at the vertex of the area lying on the line M gives the optimum values of x_1^* and x_2^*.

In practice, pets are usually not fed in such a calculating way, but for farmers such calculations can make a big difference to their profits. In more than two variables the simple geometrical method has to be replaced by less intuitive, and less educative, algebraic recipes.

The same arguments applied to the human diet would turn us all into vegetarians—because of the high cost of animal protein (meat!) which is produced by the animals from vegetable protein with an efficiency of less than 40 per cent. Since few of us are in fact vegetarians it is clear that in choosing our own diet either the function we are optimizing, or the constraints, are significantly different. It is of real interest to try and decide what they might be. One of the earliest British commercial computers, the Leo developed by J. Lyons and Company Limited, was used for this sort of planning in the restaurant business.

We could equally well treat the rabbit run problem as a non-linear version of the same type. Instead of using the constraint to eliminate L from the expression for the area, we could find the values of the two variables W and L which give the maximum area $A = WL$ subject to the constraint on the total length, which is the straight line in Figure 4.5. The curves of constant A are the rectangular hyperbolae; A is a constant on each curve and increases in the direction of the arrow. Clearly the largest value of A compatible with the constraint comes when the curve just touches the line—by symmetry, or geometry this is where $W = L = 3$.

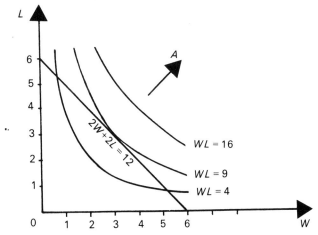

FIGURE 4.5

This is a simple illustration of a *non-linear* programming approach.

The greater clarity of the graphical arguments over the algebraic is, for most people, generally true provided they are sufficiently practised in the drawing and interpretation of the particular kind of graph involved. Unfortunately for practical purposes, such intuitive grasp is limited to two, or at most three, dimensions; however, many advanced algebraic ideas generalize to *n* dimensions so some intuitive grasp carries over.

Notice that we have used two completely different types of graph in this section. Figure 4.3 represents a function of one variable and explicitly shows its value as the vertical height of the curve above the *W* axis. Figures 4.4 and 4.5 are essentially contour plots of functions of two variables—to show the height here we should have to make a three-dimensional model, representing the function by a surface whose height above the co-ordinate plane is equal to the value of the function. It is well worth doing this from time to time to aid three-dimensional visualization.

Critical path analysis

We conclude the section by describing a particular technique of optimization by the enumeration of possibilities, which has come into the curriculum in recent years and is of considerable practical importance in the scheduling of jobs in industry. It assumes that the whole task can be broken down into a series of steps each taking a known time, some of them depending on the earlier completion of others. The method consists of an algorithm for drawing a diagram describing the necessary sequencing of steps, and for deducing from it both the minimum time that the job can take and those tasks lying in the *critical path* through the diagram; any delay in these particular steps will delay the completion of the whole task.

We illustrate the method with a simple example from *The School Mathematics Project Book* 4, page 138. Figure 4.6 shows the planning network for the preparation of a dinner consisting of minced toad-in-the-

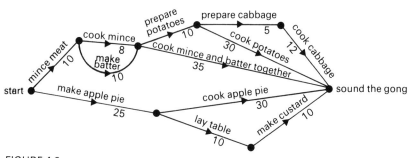

FIGURE 4.6

hole, potatoes and cabbage, apple pie and custard. The network diagram is a model of the processes involved; here we shall be content to comment on some of the features. Although towards the end five activities are happening simultaneously, four of them are cooking processes which can go on without human intervention. (One may question the absence of any time for getting the food ready to serve at the end.) However, two people are required at the beginning since the mincing of the meat and the making of the apple pie are shown as simultaneous and not sequential processes. Indeed two people seem to be involved until the cabbage has been prepared.

The critical path through the network is the one which gives the longest total time from start to finish, in this case 60 minutes. It is here that any effort to speed up the preparation of the dinner should be concentrated, though care must be taken to ensure that any saving of time does not make another path critical—for example if the potatoes were cut up small so they cooked in 10 minutes less, only 5 minutes would be saved on the total time, since two other paths add up to 55 minutes. A complete restructuring of the diagram might, in principle, offer further possibilities for saving; though in this case the making and cooking of the apple pie seem to imply an absolute minimum time of 55 minutes, there is in fact no reason why the apple pie should be ready at the beginning of the meal.

The methods described so far may seem unnecessarily elaborate for this simple illustration, and not sufficiently systematic for larger, more complex problems. It needs an explicit algorithm for identifying a critical path in a complicated diagram. One possible approach is the 'two-sweep method'. Figure 4.7 shows a network for some unspecified task; the dotted arrows, sometimes called 'dummy activities', require us to wait at

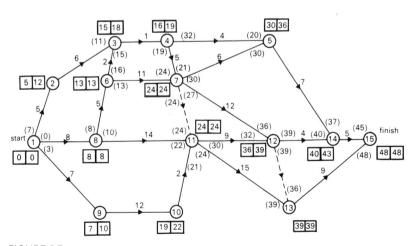

FIGURE 4.7

their terminal node until all paths leading to their initial node have been completed. The two numbers in the rectangular boxes record earliest and latest possible times of arrival at the node in the following sense. The left-hand number gives the longest time for reaching a node by any path from the start; we find it in a sweep from the left through the diagram. The right-hand number at each node gives the earliest possible time of leaving the node that will enable us to get to the finish 'on time'. We find it by sweeping back from the right of the diagram. *The critical path joins those nodes for which the two times are equal.*

An able group can learn a lot in trying to discover a systematic way for handling critical path analysis. One of its key advantages in industrial practice is that it can be computerized in a straightforward way; this too will offer an interesting challenge to able pupils.

4.3   Calculators and computers*

Micro-electronics in the curriculum

Among the newer mathematical techniques which have begun to assume importance in schools, none is more significant than those depending on the new technology of electronic calculators and computers; since they are only now becoming part of the equipment of most mathematics

---

* This section was written with Diana Burkhardt.

teachers, we shall discuss them in some detail, trying to indicate the range of possibilities that they seem to offer. Although computer facilities have been available in some schools in various parts of the country for more than ten years, the recently arrived hand-held calculators probably present greater opportunities and more urgent problems—problems because the decision of most GCE boards to allow calculators in their examinations faces the teacher with the task of integrating them into his teaching as best he can, with rather little experience or material to draw on. We shall therefore begin this section with a discussion of the possible roles of the calculator, going on later to discuss the possibilities that exist for using computers in mathematics teaching.

The dividing line between calculators and computers is now completely blurred—programmable calculators with storage and printing facilities merge in both price and capability into the bottom of the microcomputer range, which in turn leads smoothly up to the largest multi-user computer configuration. However, for our purposes the distinction is still clear. By a calculator we shall mean a cheap, personal hand-held electronic calculator—a piece of equipment which most children can be expected to own, since it costs less than a pair of football boots. This present situation, like most others in this chapter, may change over the next few years—further electronic facilities are likely to become available at minimal prices, but any mechanical sophistication such as printing or automatic magnetic card reading, will probably remain expensive.

Calculators in mathematics teaching

The absorption of calculators into the teaching of mathematics and other subjects is a major task of curriculum development which is bound to take a decade or so to complete, even as a first attempt. Here we can only describe some of the answers that early research work has suggested, give a few references, and urge teachers to begin their own investigations. Although in the context of applications of mathematics, we shall mainly be interested in the obvious uses of the calculator as a powerful aid to arithmetic, we shall begin by outlining the wide range of possibilities for the use of calculators as a *teaching* resource. Further details and samples of teaching materials and pupils' responses can be found in:

1 *A Calculator Experiment in a Primary School*, Alan Bell, Hugh Burkhardt, Alistair McIntosh and Gillian Moore, The Shell Centre for Mathematical Education, University of Nottingham, 1978.

2 *Computer-Assisted Mathematics Programme*, David Johnson and

others, Scott, Foresman & Co., 1969–72.
3 *Arithmetic Teacher*, December 1976.
4 *Calculators in Schools*, from The School Mathematics Project, Westfield College, London.
5 *Calculators*, available from Leapfrogs, Coldharbour, Newton St Cyres, Exeter, Devon.

Although (1) relates to primary schools, the observations are relevant to the learning of basic mathematical skills at any age. Unfortunately the massive store of teaching material contained in (2) is not now readily available—though it was developed for the computer, much of it can be used with modern hand-held calculators.

SMP have published four supplementary booklets related to their 'with-calculators' option at O-Level; other pupil material, from the Durham Schools Council Secondary Project and the Shell Centre/Leicestershire Primary Project should be published in 1982. An up-to-date list of sources is available from the Mathematical Association.

There is a rich variety of uses for calculators in the teaching of basic mathematical skills—we believe their role will be particularly important in the upper end of primary school, the lower end of secondary school and for remedial teaching. The Shell Centre for Mathematical Education has conducted an experiment in a primary school (1) which suggests that the opportunities for using the calculators as a teaching aid are extensive and that the commonly expressed fears that they will undermine the children's own numerical skills are unjustified. In particular, far from undermining basic arithmetic, the calculator seems to encourage and help children in developing their own number skills—this emerges from several other research studies as well. The calculator was welcomed throughout the ability range—to the bright it gave stimulation and the power to extend their mathematical exploration, while for the less able it was also an infinitely patient and non-threatening check to their misunderstandings of principles, which were revealed in a rather clear way.

Different ways of using the calculator include forecasting and checking calculations, generating examples and generalizing from them, devising games to extend skills in mental arithmetic and using results from the calculator to provoke the study of new concepts.

The widespread availability of calculators will mean that some personal number skills will no longer prove cost effective. This is an area needing careful study and experiment; a start has been made in various countries, particularly Sweden and the USA. To us it seems certain that log tables are outmoded and that fluency in long multiplication and division need no longer be taught. At the other extreme we believe that the three basic skills of a good understanding of number and place value, facility in single-digit arithmetic, and the ability to check and estimate answers, will still be worth the effort of achievement for most children. Between these limits, the best balance of curriculum objectives for each ability level remains to be discovered.

The calculator as a calculating aid

We now return to the more obvious uses of calculators in increasing the speed, power and accuracy of the pupil (and the teacher) in arithmetic calculation and function evaluation. For applications of mathematics this is a tremendous benefit in at least the following ways.

1 More realistic problems with realistic (long and nasty) numbers can now be tackled and the straightjacket of book problems with only simple numbers can be removed. For example, rooms can be 4.23 by 3.57 metres with the chimney breast 0.53 by 2.10 metres and carpets can come in widths of 0.69 metres.

2 Since the calculations themselves demand less effort from the pupil, more attention can be given to the strategy—how to approach the problem and *which* calculation to perform at each stage. For example, in problems of proportional reasoning it is found that many pupils can manage doubling and halving but that scaling by other factors is much more difficult. Can the calculator help here?

3 The increased speed and power allows new, more extended types of problems to be introduced. For example, if inflation is estimated at 10%, 22% 18%, 12% and 8% in 5 successive years, and a car bought for £850 at the beginning of the five year period is sold for £800 at the end, by how much has the value of the pound dropped, and by what fraction has the car depreciated over the period?

4 Data collected by observation and measurement in or out of the classroom can be handled much more freely. For example, a traffic survey of cars passing the school can measure the average number of cars per minute, distribution of the intervals between cars, types and ages of cars, etc. By treating bigger samples with the same effort the calculator reduces the drudgery or gives statistically better results.

5 The exploration of particular cases, which is an essential skill of the applied mathematician, is much more feasible—so algebraic models can be based on, or tested by, numerical studies of particular situations. A general form of the procedures employed, either at the flowchart level, or as an algebraic expression, or preferably both, will remain the ultimate objective of such

exploration. When combined with practice in tabulating algebraic expressions for a range of values of the variables, this may help to overcome the threshold difficulties of algebra.

Examples that could be investigated include:

- home central heating. Work it out for one wall, then one room, then your house, so that gradually the general principles emerge.
- planning a walking holiday, or a car journey. Times can be estimated from distances and assumed speeds. Money to be spent on essentials can be calculated.
- exponential growth. See how 5 per cent per year builds up over a period, then 10 per cent, and so on.
- the effect of compounding interest half-yearly, monthly, daily (and so discover e). Learn that over a period these changes amount to a change in the effective interest rate, nothing more.

To pursue this further, look at the effect of compounding interest more frequently. Compounding n times per year requires an annual interest rate of i such that

$$\left(1+\frac{i}{n}\right)^n = 1+I$$

where I is the rate compounding annually. Thus:

$$i = n\left[(1+I)^{\frac{1}{n}} - 1\right].$$

Evaluating such an expression is possible with log tables but very tedious; on a scientific calculator it is straightforward although not elementary, perhaps involving:

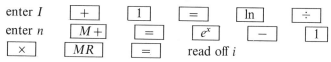

enter I $\boxed{+}$ $\boxed{1}$ $\boxed{=}$ $\boxed{\ln}$ $\boxed{\div}$

enter n $\boxed{M+}$ $\boxed{=}$ $\boxed{e^x}$ $\boxed{-}$ $\boxed{1}$

$\boxed{\times}$ $\boxed{MR}$ $\boxed{=}$ read off i

6 An extension of this type of exploration is the 'trial and error' approach to testing a hunch about how to tackle a problem. Calculators are easy and quick to use so you can try a few numbers.

- Find the quickest route from a point in a rectangular swimming pool to a far corner, given swimming and running speeds.
- Find the maximum area of a sheep pen that can be made with seven straight line hurdles at the side of a field.

Many of these opportunities amount to allowing greater demands on the more sophisticated skills of the pupil in problem formulation and solution, because the calculator takes over the routine (but not easy) arithmetic. We must be careful to ensure that any such extra demands are reasonable.

No doubt more material will emerge but it is going to take time and a lot of experiment before a considered calculator-aware curriculum is devised. The Schools Calculators Working Party provides a channel of communication between those active in such exploration; it can be contacted through The Shell Centre for Mathematical Education, University of Nottingham, which provides its administrative support.

Computers in mathematics teaching

This is not the place for a review of the present state of computer education in schools—an interesting worldwide summary is given in *Informatics and Mathematics in Secondary Schools*, edited by D.C. Johnson and J.D. Tinsley, North-Holland, 1978. We can particularly recommend the interested teacher to look at the journal *Computer Education*. The shortage of trained specialist teachers in this rapidly expanding field means that these publications inevitably concentrate on teaching computing, including non-numerical information processing as well as numerical computation. Computing skills are useful far beyond the mathematics class but only in the way that mathematics itself should be.

The essential facilities in which the computer goes beyond the calculator are as follows.

1 Many numbers can be stored, forming a basis for much more extensive investigations. Non-numeric data can be handled too.
2 A stored program allows the automatic repetition of calculations on different cases, or by iteration with continuous improvement.
3 The visual display and/or printer allow far more extensive and readable output of numbers or graphs, which is further enhanced by automatic graph plotters. Magnetic tapes or disks allow the transfer of huge amounts of data.

The storage of data is perhaps the key to the most important educational benefits of the computer. It not only allows the student to build up his own data collection in a form which can be readily handled, but allows him to use data bases that have been compiled on particular topics. Such collections may contain textual information, such as names, as well as numbers.

The price of using some computer power in a school is still high enough to deter most teachers from becoming involved. Problems lie in a number of different directions, quite apart from the initial financial outlay by the school authorities. Recent progress on this front, envisaging at least one microcomputer in each secondary school by 1982, must be viewed in the perspective of vastly greater potential demand.

1 A lot of energy is needed to get the school equipped even with simple facilities.

2 Computers are complex machines and ingenuity may be needed to keep the system running. The standards of reliability set by a piece of chalk, a blackboard and a duster are still beyond the reach of most other teaching systems.

3 The management problems of the class, the machines and program processing are an extra load that only energetic teachers are prepared to contemplate.

All this helps to explain the difficulty of the transition from the pioneering stage, in which computers are the concern of only the enthusiastic teachers, to widespread computer education for all.

However, the pace of technological change is such that:

1 We have now reached the point where, in 1981, a powerful self-contained computer with:
   - thousands of memory locations
   - the BASIC programming language
   - thousands of computations per second
   - keyboard and visual display
   - magnetic tape casette or disk input/output
   is available in classrooms for about £200.

2 There is little doubt that within a few years similar computing power and storage, with a more limited visual display and no tape or disk, will be available in the £20 region as a personal aid to the pupil. It will be easily linkable to a bigger computer for exchanging programs and data.

The educational possibilities of today's microcomputers should be seen in these terms; we are only just beginning to explore them.

Let us look at what the computer can bring to the mathematics classroom.

1 Programming gives pupils experience of algorithm design and execution, and this can be tested by the computer. This is an extension of the flowchart work that has provided a language for precise, logical thinking which seems to be acceptable to most children. The preparation of programs in the simple BASIC language, sitting at the keyboard with immediate checks on the syntax, is a highly addictive activity* that most children, and adults, enjoy (the initial fear hurdle is easily overcome with proper teaching). In programming mastery must, and can, be achieved.

2 Modern computers including micros, have facilities for a wide range of graphical output. They allow the display on the visual display unit (VDU) of:

---

* We incline to the view that early innoculation against this disease is advisable in order to avoid too virulent outbreaks at a later age!

  – graphs of functions or phenomena as a function of various parameters
  – transformations of graphs, to examine details, to watch the function change in time, etc.
  – lists of words or other information, mathematical definitions, etc.
  – patterns

All these things may be explored dynamically by the user if he so wants.

A versatile graphics facility raises the possibility of intensive training in translation skills (see page 115). For example:

  a *From graphs to words:* a set of stored curves with labelled axes representing real phenomena can be interpreted by the student, for example:
    – car speed v. distance round a racing track
    – pulse rate and breathing rate v. time in various circumstances
    – height v. age for boys and girls
  b *From graphs to algebraic expressions:* a curve is shown and a student invited to choose the best fit from a set of algebraic expressions with adjustable parameters. His choice is displayed and he can correct and improve it.
  c *From algebraic expressions to graphs* (sketching curves): the student inputs points onto a screen to represent a curve whose algebraic expression should have been given; he then gets the computer to sketch it as a check on form and accuracy.

Work has begun in some of these areas.

3 Data which has been compiled by others or by the class over a period of time. Thousands of items of data can be stored in the computer which can then be questioned. For example:

  a Football league results for last season:
    – What happened to Manchester United through the season?
    – How many draws and score draws were there each week, and so on, and on average?
    – How good a prediction of the result does the half-time score give?
    – Which teams attract the biggest crowds and how much difference does this make to results?
  b Figures of births, marriages, divorces and deaths for the last hundred years:
    – What happened to the birth rate? Why?
    – What effect will it have on the population?

- How many children will there be in secondary schools next year? In 1984? How many teachers will there be?
- Are people marrying younger, more often?
- At what age do most people die? Why is it changing and by how much?

The development of data bases related to other school subjects (for example, census information relevant to geography or to history) is gradually spreading—in the long run this implies an extension of the service role of mathematics and computing, whether they are organized separately or together.

4 *Numerical methods* and other algorithmic approaches to solving mathematical problems will have an enhanced role. The availability of computing power alters the balance of judgment as to what is an efficient way to tackle a mathematical problem. The drudgery of arithmetic has in the past put an almost total emphasis on algebraic methods giving solutions in closed form, but there are many problems (e.g. roots of a cubic equation) where a numerical approach is quicker. There are many more (area under the curve of a given arbitrary function) where numerical solution is the only general method. This re-thinking of the mathematics curriculum may prove radical and fundamental, but it will be slow.

Program packages

There is little doubt that a major future use of computers in the classroom will involve 'teaching packages'—programs on magnetic tape or disk that are loaded into the computer in a simple, standard way and which provide the teacher, or the individual student, with an aid to the teaching and learning of a particular point or topic. A few such packages exist now but there is little doubt that mathematics and its applications can benefit and that provided they are easy and profitable to use, computers will have a place in its teaching.

Packages can be used for a wide variety of learning activities.

Generating examples and exercises
The computer responds to the pupil, marking answers and providing properly adjusted progressive teaching and testing of the mastery of routine skills. For simple arithmetic skills such a package is available on special hand calculators (e.g. 'the little professor') with perhaps a thousand examples at four levels of difficulty for each arithmetic operation $(+ - \times \div)$; a microcomputer could have many such packages, with more versatile feedback and reinforcement exercises.

Illustrating mathematical points

The representation of a function by a cartesian graph is not easy to grasp (see page 117). The graphical facilities of the microcomputer allow the student to illustrate this and other graphical representations on a sufficient range of cases to make the principles familiar. The labour of plotting points by hand has often meant that in the past this kind of generalization (and others) has been forced on the pupil before it seems inevitable or cost-effective.

Simulating practical situations

The role of the computer in simulations of various kinds has already been mentioned. Since the carrying out of experiments is frequently outside the experience, and even the ambition, of mathematics teachers, a program package which simulates such results is of value, e.g.:

1 The famous, or notorious, moon landing game, which is available on most microcomputers now, requires the user to control the burning of retro-rocket fuel so that the rocket lands 'softly'. It can be tackled at every level from enjoying the crash to landing with minimum expenditure of fuel, and it calls upon a range of mathematical skills including the ability to model.

2 The simulation of situations with a random element by a computer allows a much more rapid acquisition of data. The tossing of a coin ten times may give 'heads' from 0 to 10 times—it takes perhaps 30 seconds. To get a good idea of how likely each of these possibilities is, one must repeat the experiment a hundred, or better several thousand, times and this takes a lot of time and organization even with a class. A computer can simulate this and display the histogram of the proportion of time each number of 'heads' has occurred, and the class see it approach the binomial distribution. What about sequences of 20 tosses? In this way a feeling for distributions—binomial, normal, Poisson—can be built up long before the mathematical machinery for a derivation is available. 'Statistical intuition' takes time to develop— experiments are important and simulations have an undoubted role to play.

In many simulation packages the student is asked to contribute something to the model the package uses to generate the data; usually it gives only the values of some parameters (perhaps a number of coin tosses in each sequence, the number of sequences required or even the bias on the coin), though packages can be written which ask the student to provide the model itself, i.e. to supply functions connecting the key variables.

1 At a very simple level a graph-plotting package asks for a statement giving the function to be plotted.

LET $Y = X * X + 1$

or

LET $Y = 0.5 * SIN(X) * EXP(-2 * X)$

2 A car acceleration program may ask for maximum power P watts as a function of engine revolutions W per second.

LET $P = 5\,000\,000 * W/(10\,000 + W \uparrow 2)$

Computer-managed learning

Packages of a different kind can be used to manage routine aspects of teaching. Hertfordshire for example has developed a mathematics scheme where the computer marks tests, keeps records of individual students and suggests suitable remedial programmes.

4.4   Statistics

A Cautionary tale—the breakdown of a model

A student very much wants to own a car. He has saved some money and he can buy one that is on offer. He calculates what it will cost to run it. The tax, insurance, petrol and garaging costs are straightforward to estimate; servicing and repair costs he discovers from a table of average costs in a motoring magazine, which gives for his type and age of car £300 per year. This total running cost, he decides, is just about acceptable within his income, so he buys the car.

He is not too surprised that there are a number of repairs needed quite soon—new brake linings and an exhaust system are the main items in a bill of £120. However, within three months a very distressing banging noise inside the engine requires a complete engine overhaul (£200) and an MoT inspection demands chassis welding (£140). This he cannot afford, at least at the time, and the car is taken off the road.

What went wrong? The calculations he made were sensible but one item, the repair costs, was much larger than the average given by a survey of owners. His deterministic model, though sensible, is not good enough—he needs statistics, from several points of view.

First it is obvious that repair costs are subject to large variations which cannot, at least in practice, be predicted and that this element of randomness is important in this case because the spread of fairly likely costs is wide enough to make a significant difference—i.e. between his being able to afford the car or not. (If he had a larger income, the average cost might well be sufficient, since he could absorb fluctuations without serious embarrassment.) So he needs a statistical, (or *stochastic*) model, which takes possible fluctuations into account.

There are other more subtle matters which are important and which come under the umbrella of statistics. The average costs given previously were found by a motoring organization sending a questionnaire to a sample of its members. Was the pattern of car use in this group close enough to our student's in mileage per month and so on? Do people accurately report their costs or do they under-report. Does an old car tend to need more repairs when newly acquired? Do members of motoring organizations take better care of their cars than others? And so on. This sort of reasoning skill is of importance to everyone; it is the essence of statistics. It is not easy to acquire and has not perhaps been explicitly taught in the past. In this chapter we shall discuss how it can be done. It does of course involve modelling of both sorts (see page 106)—descriptive modelling of the data and some analytical modelling of its underlying structure; in some important ways it is more subtle than straightforward deterministic modelling and this needs some further discussion.

The nature of stochastic modelling

There are a number of factors which distinguish statistics from the rest of applied mathematics, and make its teaching significantly more difficult. (Some statisticians in frustration at the misunderstanding of their fellow mathematicians claim that their's is 'a different subject'—this seems a sterile argument about definitions. We believe that all mathematics has a lot to learn from recent teaching developments in statistics.) Although it has been claimed that statistics is 'the study of data', it is specifically concerned with building models of situations where there is a significant amount of randomness or of uncontrolled variation of relevant variables. *This means that the predictions of the model cannot be checked by a simple experiment or observation as they can in a deterministic situation.* So for example, the standard falling body model predicts a fall from rest of 5 metres in the first second—a flash photograph can confirm or refute this. On the other hand, a probabilistic model says that the average number of 'heads' in tossing a coin is one-half the number of tosses—a different proportion is likely to be found in any experiment, but this does not refute the model. Rather, one must *go back to the model* and predict the expected variation of this proportion. This again will not agree with experiment but will lead to further questions to which the probabilistic model can provide answers but with similar limitations.

This much more indirect and sophisticated relation between the model and the practical situation is the essential extra difficulty in teaching statistics. It is probably too difficult a concept for most children even at 16; in which case the subject must be approached on an intuitive and heuristic basis at this level. In particular, the links between the statistical procedures and the probabilistic models may only be very partially made. A lot of work on this difficult problem of curriculum development has been carried out in many countries, but few people would claim that a satisfactory solution has been found.

On the other hand, situations to which statistics is applied are often approached from a much more realistic and balanced point of view than is common in conventional applied mathematics—the variation of the weight of biscuits in a packet or the distribution of marks in an examination may seem to be of rather more interest to most people than the timings of a cyclist on a journey, or simple interest on stocks and shares. The recent interest in more realistic and relevant applications of mathematics in general, to which this book is largely devoted, certainly owes something to the example given in statistics courses. Much attention is also paid in most statistics courses to developing skills in the selection and collection of data, its presentation in tables or graphs of various sorts, and the translation of information from one form to another; we have stressed that these are all skills which are equally important in all areas of realistic applied mathematics.

However, in tune with the current general trend, a strong movement has developed in the last few years to try to make school statistics more realistic and more useful to those who learn it. Of particular interest is the Schools Council Project on Statistical Education (POSE), which is based at the University of Sheffield but whose material is being developed and tested in 55 schools around the country, with many others involved to a lesser extent. The background information that has been collected and published in their project papers is most valuable, while the classroom material they are developing seems extremely promising; we are grateful to the Project and its Director, Peter Holmes, for permission to quote from these sources. Our enthusiasm arises partly from the similarity of their approach to that which we are advocating—they draw applications from realistic situations of potential interest to the pupils, while simultaneously building up a range of techniques for constructing models which are firmly based on the empirical data. They have interesting examples for many areas, a few of which are Action problems for the pupils as shown in the table opposite.

Looking at real data

Statistics courses have always been concerned with the representation of data; there has usually been some encouragement for the collection of data by pupils but, since this takes time and causes disruption, it is not usually emphasized. Less common is any attempt to develop the critical

Area	Examples
Games and gambling	'pools', horses, cards, sport in general
Insurance	life, motor, theft, fire
Health and Medicine	vaccine, drug trial, epidemics, smoking, fluoride and teeth, diagnosis, screening, drugs
Biology	genetics, heredity, germination, conservation and pollution, drinking and driving—reaction times

appraisal of data which is the essence of statistics and fundamental to the drawing of reasonable inferences. The POSE material is firmly based on teaching children to do just this. An early unit begins with the collection and representation of data.

A Leisure Time

A1 Some Questions

We all have some time to spend as we want. In the future we expect people to have even more spare time. We want to make the best use of this leisure time.

Think about these questions:

What do you do in your spare time?

How long do you spend watching television?

Which programmes do you prefer?

What games do you play?

What kind of books do you read?

a Write down some things you do in your spare time.

b Who would like to know which books you read?

c Why?

On page R1 there is a section called 'My Diary'. During the next seven days note down the amount of time you spend each day watching television and the amount of time you spend reading. Do not count time reading schoolwork.

A2 Today

Spare time is leisure time. Let's see how you spend today.

On page R1 you will find a clockface headed 'Today'. The clock face is to show how you will spend this afternoon and evening (from midday to midnight).

a Start at midday and show how you will spend the 12 hours until midnight today. Your clockface should look a bit like the one drawn in Figure 1.

Figure 1

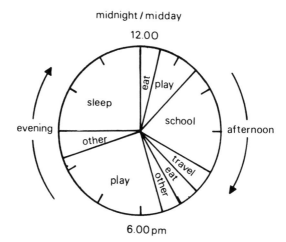

b Make a table like Table 1. Put in your figures for today.

Table 1 My time today

Activity	Time in hours
Eat	$\frac{1}{2} + \frac{1}{2} = 1$
Play	$1 + 3 = 4$
School	$2\frac{1}{2} = 2\frac{1}{2}$
Sleep	$3 = 3$
Other (including travel)	$\frac{1}{2} + \frac{1}{2} + \frac{1}{2} = 1\frac{1}{2}$

c   **Complete the clockface on page R1 to show what you did last Sunday (from midday to midnight).**

d   **Make a table like Table 1. Put in your figures for last Sunday.**

The times from Table 1 can be put on to a circle. There is now only one section for each activity. This circle represents the 12 hours today from midday to midnight, but it is *not* a clockface. It is called a PIE CHART.

Here is the pie chart for our Table 1 figures.

Figure 2   *Time spent today*

360 degrees represent 12 hours
  30 degrees represent 1 hour

e   **Complete the two pie charts on page R1. These show how you spent your time today and last Sunday.**

f   **How do your two pie charts differ?**

*g   **Find out how a working adult you know spent his time today.**

    Draw a pie chart for him.

    How is it different from yours?

    Did he have less leisure time than you?

B Leisure Activity

B1 What Do You Do?

a Complete the questionnaire on page R1.

b Record the class results on pages R2 and R3.

B2 Favourite Television Programmes

Many people like to watch television. The favourite television programmes of Class 1b of Creektown School are shown in Table 2. (The title 'other' includes all programmes that had only one vote.)

Table 2   *Favourite television programmes of Class 1b*

Title	No. of pupils
Match of the Day	12
Bionic Woman	8
Batman	4
Dr Who	5
Other	3
	32

They showed these figures on a bar chart (Figure 3). Notice that there is a title, both axes are labelled and the bars are all the same width. The gaps between the bars are also the same width.

Figure 3   *Favourite television programmes of Class 1b*

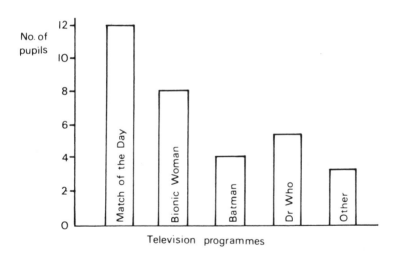

Match of the Day is the favourite television programme of Class 1b. For them, *Match of the Day* is the MODE.

a Draw a bar chart to show the favourite TV programmes of your class.

Later the difficulties of collecting reliable information through questionnaires are treated in this way:

A   School Rules

Schools have rules. Pupils who break rules are punished. Some teachers use the cane. This is called corporal punishment. Some people agree with corporal punishment. Others do not. Read this newspaper cutting.

SCOTS PUPILS VOTE FOR STRAP

Two out of three secondary school pupils in Scotland think that teachers should be allowed to use the strap. This figure is based on a survey made last summer. A questionnaire on a wide range of rules was given to a large sample of Scottish children.

Pupils said that the strap was needed to correct bad behaviour. It was needed for control in the classroom. Pupils agreed with most school rules except those on clothes and appearance. 43 percent of the girls and 16 percent of the boys thought that pupils should be able

It shows what 1000 Scottish children think.

What are your school rules?

How are pupils in your school punished?

Are the punishments fair?

One way to find the answers to these questions is to ask everyone in the class. This is a SURVEY.

We use a QUESTIONNAIRE to make sure that everyone answers the same questions.

Here are two questions for a survey of school rules.

'How many school rules have you broken?'

'Do you think corporal punishment should be allowed?'

a Why is the first question difficult to answer?

b Write three more questions for the survey.

Exchange your questions with a friend.

c Try to answer your friend's questions.

The detailed illustrations include alternative sample questionnaires leading finally to some guidelines for designing a questionnaire.

The misleading presentation of data is analyzed in a later unit—Phoney Figures.

In January 1979 Catchunks published the advertisement shown in Figure 1.

Figure 1   First advertisement for Catchunks

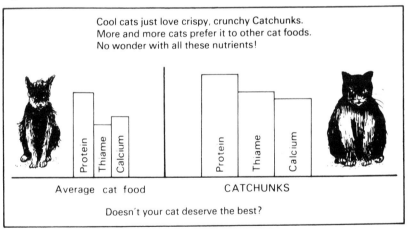

a How does this advertisement mislead?

The POSE material is published by W. Foulsham & Co. Ltd. We have given several illustrations of the POSE approach to teaching the handling of real data because we feel that they provide a concrete example of how these difficult skills can be taught. Most of them are equally relevant to the handling of data from other areas of applied mathematics where random variation may not be important. They are an essential element in the formulation and, particularly, the validation stages of building and using mathematical models.

Probability

The theory of probability provides the mathematical models on the basis of which statistical inferences are drawn; it also gives an abundance of exercises in combinatorial thinking and algebraic manipulation. As we have said, however, these processes of inference can really only be approached intuitively at this level and, even with the abler students, only the simplest models used in practice can be derived from first principles.

The major area of everyday applicability of probability ideas is undoubtedly gambling, and the ability of the keen punter to 'reckon odds' predictably astounds those who have tried to impart these difficult ideas in a less strongly motivated, less concrete situation.

Traditionally there has been a strong emphasis on situations of equal likelihood (pennies, dice, etc.) where the additive and multiplicative laws of probability can be demonstrated and practised by the pure counting arguments of combinatorics. While this approach may be adequate for most table games, the horse, and life itself, undoubtedly require the ideas of a relative frequency and its limit, as with loaded dice, and of the subjective probability of individual or group opinion. (Subjective probability has never been taught systematically at this level.) From the teaching point of view, probability has all the attractions of mathematics—clean, clearly specified assumptions, calculational procedures and answers. The modelling problems cannot be avoided entirely, but with pennies, cards, dice and balls in bags they can be adequately stylized, leaving a subject area a mathematician can feel happy in, provided he can be protected from the messy uncertainties of real data and inference from it. However, in our view the student must learn to sail the sea outside the harbour of mathematics, so the statistics curriculum should contain both aspects, with an emphasis on data. The probability theory should be taught and learnt in such a way as to build up stochastic intuition—useful in real problems. This will often require simulation with large samples and, since the tossing of pennies and the throwing of dice is a slow business with a limited choice of *a priori* probabilities, we again point to an important role for microcomputers in this area. The first few cases should surely be done with concrete experiment by the class, but the computer can then swiftly enrich the range of cases studied—we need good, graphic educational software to show this in a way that avoids breaking the link with reality.

Games with 'incentives' have surely a role to play here more than elsewhere:

1 rolling two dice:
A wins if total is 2, 3, 4, 5, 10, 11, 12
B wins if total is 6, 7, 8, 9
First to get a 6 chooses to be A or B; most will unwisely choose A and gradually learn about non-uniform distributions. The computer can help build on this lesson with more 'dice'.

2 Desmond Bagley's book *The Spoilers* describes a team of adventurers out to destroy a heroin 'ring'. One of them continually challenges another to probability-based games which seem to be fair, and equally regularly wins. In one example each player chooses one face of a coin and they simultaneously show their choice to each other. Player A, the instigator, offers the following odds:

A's choice	B's choice	A wins	B wins
Heads	Heads	£20	
Heads	Tails		£10
Tails	Heads		£30
Tails	Tails	£20	

The game appears to be fair but by choosing heads rather more than half the time player A can ensure that he has an advantage. This is not obvious. However, if A chooses heads randomly with probability a, and B with probability b, then on average A will win an amount $W = 20ab - 10a(1-b) - 30(1-a)b + 20(1-a)$ $(1-b)$ £s per time. Now a and b must lie between 0 and 1; $a = b = \frac{1}{2}$ gives $W = 0$. However, for $6/10 < a < 2/3$, W is always positive for any b (try $a = 5/8$, or put $W = 0$ and plot a against b).

Games which require pupils to commit themselves to a prediction in the way described above seem to build up stochastic intuition, a feeling for randomness, more rapidly than academic exercises, probably through the more vivid memory of experiences of success and failure. There are enough surprising results in probabilistic situations to make this a rich vein of teaching.

A curious example is the birthday problem. What are the chances that two children in the class have the same birthday? Or what is the likelihood that two cars in the next 50 to pass have the same 3-digit number on their number plates?

Most of us are surprised how quickly the product

$$1 \times \frac{364}{365} \times \frac{363}{365} \times \ldots \frac{365-n}{365}$$

which gives the probability that $n+1$ children all have different birthdays, gets smaller. Later some children may be able to see that it can be approximated by

$$\left(1-\frac{1}{365}\right)^{0+1+2+\ldots+n} = \left(1-\frac{1}{365}\right)^{\frac{n^2}{2}}$$

allowing a shorter calculation.

A practical example is crossing the road. Do cars flow evenly spaced along a road? If not, how do they fluctuate? As well as collecting data on this interesting problem, pupils can tackle it from the point of view of probability theory at various levels. Perhaps the simplest approach is to assume that ten cars a minute are passing by (an average of one every six seconds), to consider time intervals of one second, rolling dice and assuming a car passes if a six is thrown. Combining results in the whole class and assuming that the organization hazards of this sort of activity have not loomed too large, a fair feeling for the distribution of time intervals between successive cars will emerge. The importance of these fluctuations in enabling people to cross busy roads can be discussed, as can the deviations from this random model produced by traffic lights, slow lorries, and even perhaps speed limits.

The essentials of a curriculum

We have argued that an understanding of statistical methods for pupils in this age range may best be gained by building intuitively on their experience of more or less real situations. This modelling approach which, we have said, is important for developing skill in applying mathematics of all sorts, seems almost unavoidable in statistics at this introductory stage. It seems to us that there are certain essential skills which a pupil must acquire if statistics is to be of any coherent use to him and that it is worthwhile trying to identify what these are.

The skills in the collection, organization and interpretation of data that we have mentioned must clearly be included; these are represented in most statistics courses in the 11–16 age range, indeed they often seem over-represented, but there is need for more attention to be paid to the difficulties of handling real data. At the theoretical end of the spectrum of skills, probability calculations provide mathematical challenge and interest at least for the able student; their difficulty is such that only very simple practical situations are normally discussed in detail. It is hard to see a great deal of room for improvement here except perhaps through the increased use of simulations using microcomputers.

Between the data itself and the probability models lie the skills in looking at the data and drawing inferences from it. We would again urge that emphasis be placed on semi-quantitative intuitive approaches rather than the carrying through of numerical algorithms. The idea of a distribution, and the quite different roles of an empirical distribution obtained from data and a theoretical model distribution from probability theory, are clearly central. Much of the data will represent distributions 'with humps in'. It seems vital that students should understand the concepts that humps have a 'middle' and a 'spread', and that they also have more subtle aspects of shape which are normally refinements. The rough quantitative relations between them are also important—if the spread is very small compared to significant changes in the position of the middle, then statistical fluctuations are unimportant and the problem may usefully be handled deterministically, at least in this aspect; conversely a spread large, compared to other variables of interest, needs interpreting. Again, the importance and the difficulty of choosing sensible variables is evident.

The problems of *suitable measures* of middle and spread are less central, though the algorithms for calculating them tend to figure largely in current statistics courses and provide practice in mathematical technique. If such measures are presented as matters of convenience with a request for ideas from the pupils, they may be a source of fun, technique and understanding. Median and mode (middle one and peak?) may well appear spontaneously from a class discussion and, given calculators, the greater convenience of the mean will become evident in practice. Two measures are enough, though three are better, to illustrate the crucial point that if the differences in the results of different measures of the middle are not small compared to the spread then the situation is peculiar, and care and scepticism are good reactions. Particular features of the distribution which lead to this sort of behaviour, like long tails, can be pointed out and their interpretation discussed. Similarly, measures of spread can be pursued graphically then numerically, with the same lesson—if the difference matters it probably doesn't have a lot of meaning at this semi-quantitative level. We believe that pupils can usefully be taught the semi-quantitative ideas of distributions, peaks, middles, and spreads and their interpretation quite some time before they need be concerned with the algorithms for measuring them.

The idea of a population and a sample are equally central. It is essential to understand that observations of a sample give different results from the same observation of the population as a whole but that this limited information can be useful. Understanding the quantitative relationships between the various measures involved is much more difficult and perhaps one should often be content with a rough feeling for the order of magnitude of the fluctuations that are to be expected in a few simple but frequently occurring situations (e.g. we expect $\sqrt{N}$, where N is the expected number; this follows essentially because a random walk of N steps will on average finish at a distance of about $\sqrt{N}$ steps from the starting point).

Any model that the pupil is expected to use should be linked in his mind to a reasonable variety of standard situations which it can usefully describe.

The problems of curriculum design and implementation are discussed much more fully in the POSE handbook *Teaching Statistics 11–16*, which also describes the Project's ambitious programme for a basic secondary education in statistics. The breadth and range of statistical applications, the problems of interdisciplinary teaching of a multidisciplinary subject, and a detailed look at the syllabus from the point of view of a hierarchical ordering of syllabus content, are among the subjects discussed.

4.5 Scaling up or down

We are going to talk about scaling, which lies somewhere between a specific situation and a mathematical technique. We are concerned with the relative importance of different factors in a problem when certain variables, usually the linear dimensions, are scaled up or down by given multiples. The techniques used are proportional reasoning, or ratio, and powers. The range of actual situations and the empirical laws or principles involved are very diverse; we may discuss for example:

- Why is there a largest possible insect?
- Why is there a largest possible bird?
- Why do small cups, or pots of tea cool more quickly?

  – Why do elephants have fat legs?
  – Why is the biggest mammal a whale?
The unifying principle is that, if the linear dimensions of a body are all scaled up by a factor of, say, 3 (or x), the area is increased by 9 (x^2) and the volume by 27 (x^3). So, for example, the ratio

$$\frac{\text{surface area}}{\text{volume}} = \frac{S}{V} = \frac{9S_0}{27V_0}$$

goes *down* to one third (or $1/x$) of its previous value. Let us look in more detail at some of these examples. We shall use L for the linear size (e.g. length) and small letters for 'constants' independent of it. Some of these and other biological examples are discussed in *On being the right size* by J. B. S. Haldane in *The World of Mathematics* edited by J. R. Newman and published by Allen & Unwin.

Why is there a largest possible insect?

Insects absorb oxygen through their body surface, $S = aL^2$; they have no lungs but they use oxygen to provide energy for the mechanical work of muscles throughout their body volume, $W = wL^3$ (there is also a heat loss through the surface $H = hL^2$ but this doesn't spoil the argument; check this mathematically). So, for enough oxygen we require

$$aL^2 > k(wL^3)$$

$$L < \frac{a}{kw} \approx 10 \text{ cm for insects.}$$

All animals bigger than this have lungs with a huge surface area in many small tubes. Teachers might discuss the assumptions implicit in this argument.

Why do small cups of tea cool more quickly?

The rate of heat loss through the surface by evaporation is $H = hL^2$ per minute, while the amount of heat lost in a given fall of temperature d is proportional both to the volume and to d so

$$H = cL^3 d$$

and

$$d = \frac{h}{c} \times \frac{1}{L}$$

so that the temperature fall d in a given time goes down as L increases. More crudely—3 times the diameter, 9 times the heat loss, 27 times the heat capacity, so one-third of the temperature fall. Convective cooling is also proportional to the surface area so exactly the same argument applies.

Why do elephants have fat legs?

Again the weight $M = mL^3$ rises faster than the strength of the legs, which is proportional to the cross-sectional area of bone or muscle, $S = sL^2$. So for a scaled-up animal the legs must get proportionately thicker—hence the largest land mammals are elephants with thick legs (see also rhinoceroses and hippopotamuses). The giraffe with a very thin body exploits the other possibility. The whale is supported all over and so can be much bigger without danger of structural break up; this answers the last question in our initial list.

Dimensional scaling arguments such as these are not restricted to situations in which length is the independent variable. For example, there is a natural speed limit for cars driving on ordinary roads because the distance a driver can see is roughly independent of the speed he is travelling, while the stopping distance is proportional to its square—with the acceleration limited roughly to $g = 10 \, \text{ms}^{-2}$, both the stopping time and the mean velocity are proportional to the speed. A stopping distance d of 500 m gives a limiting speed of

$$V = \sqrt{(2gd)} = 100 \text{ms}^{-1}$$

or about 200 mph. We may like to consider the factors that make actual legal speed limits about one-third of this.

4.6   Delight in applications*

Throughout most of this book you will have found a rather practical emphasis, in which we stress the pay-off the pupil can get from learning mathematics that will help him face situations which he wants to know more about. However, the intellectual pleasure which many find in mathematics is there in abundance, in applications of all sorts, for those who seek it. Just as the contemplation of daffodils may lead to a treatise on horticulture or to a poem, so the mathematician's contemplation of the world around produces outcomes of similar diversity. This, for some of us, is part of the charm, and this safety valve of free imagination is certainly necessary for the mathematician's health; it is perhaps also essential for the development of mathematics as a connected discipline, with potential for wider applications in the future.

Here we offer a series of variegated examples; some of them come from practical problems where the power of mathematics was the prime

---

* The material in this section was contributed by Trevor Fletcher.

motivation, while others are just observations for the curious. The descriptions do not aim to present a balanced picture of the situation discussed but rather to show clearly the essential mathematical content of a particular model. Every problem should offer something to the 11–16 age group and a few have been chosen because they lead naturally to more advanced ideas as well. It may be instructive to approach many of them at different times in the school course with different techniques. Representing as they do some substantial achievements in applied mathematics, it is generally not reasonable to expect pupils to be able to formulate such models for themselves.

Knitting needles

The table on the next page shows that knitting needle sizes vary in different countries.

Plot graphs and see if you can find any relationships between the various scales. The French scale is in millimetres.'

Use of logarithmic-linear graph paper (which should be in every mathematics classroom) will show that the relation

$$E = 20 \log \frac{10}{m}$$

where E is the English number and m the metric, holds over a wide range

Knitting needle sizes		
British	French	American
14	2	0
13	—	—
12	2.50	1
11	3.00	2
10	3.25	3
—	3.50	4
9	4.00	5
8	4.50	6
7	4.75	7
6	5.00	8
5	5.50	9
4	6.00	10
3	7.00	$10\frac{1}{2}$
2	8.00	11
1	9.25	13

of values, but that it is somewhat in error for the larger sizes. Logarithmic relations occur frequently in science and technology.

Siting a telephone box

This is a problem for discussion. A number of houses are spaced at various intervals along a street. Where is the best place to put a telephone box so as to serve all of the houses as fairly as possible? What happens when the houses are situated at various places at the nodes on a square grid? (Draw examples and investigate.)

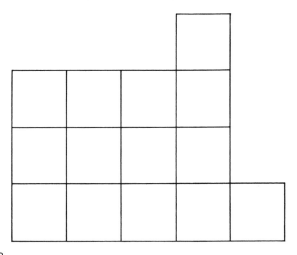

FIGURE 4.8

As a variation of the last question consider the housing estate whose street map is shown in Figure 4.8. Where is the fairest place to put a telephone box? What do you mean by fairest? If the students agree on a minimum average distance, the subsequent problems are relatively simple.

Unicursal problems

The problem of describing networks unicursally is well known and occurs in a number of textbooks. The Konigsberg Bridges problem expounded by Euler was the first and most famous of these. A related practical problem is involved in marking out a football pitch or a netball court.

If a man has to repaint the marking lines and wishes to walk the minimum distance what must he do? Students can investigate such a problem empirically and compare their solutions. It is uncertain whether any algorithm is known which provides a general solution to the problem, but a useful guiding principle is to see the problem as a matter of inserting extra links until the network is unicursal, and to try to minimize the total length of these inserted links.

Travelling salesman

The problem of the travelling salesman is a classical problem in operational research. The salesman has to visit all of the towns on the map (Figure 4.9) and he wishes to travel the minimum distance, returning eventually to his starting point. No general method of solution is known apart from enumerating possibilities. If there are n towns there are at most $\frac{1}{2}(n-1)!$ possible routes, but for even moderate values of n this number of possibilities is too large to handle on the most powerful computers.

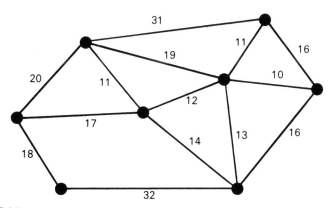

FIGURE 4.9

However, it is a good problem for the student to investigate particular cases such as provided by the accompanying diagram. Students should also be able to prove a general result—in the case when the distances are in fact airline distances (such as for a flying salesman) the optimal path never crosses itself.

Ventilation in mines

You may take the network in the last question to be a map of the galleries in a mine. The numbers indicate the minimum flow of air which is necessary for proper ventilation. You may arrange for the air to flow in either direction according to which seems most convenient. Where would you like to pump in air and to extract air, assuming that only one place can be used for each purpose? Is this a reasonable statement of the problems of ventilation which might actually be encountered in a mine? Can you reformulate the questions so as to improve the model? Can you find any methods which would apply to networks in general, and not merely to the particular network given?

Variation in athletic performance in different parts of the world

If an athlete putts the shot, then assuming the usual dynamical laws for projectiles and neglecting the resistance of the air, the distance thrown d is given by the formula

$$d = \frac{V^2 \sin 2\alpha}{2g} \left\{ 1 + \sqrt{\left(1 + \frac{2gh}{V^2 \sin 2\alpha} \right)} \right\}$$

where h is the height at which the shot is released,

   V is the speed at which the shot is released,

   α is the angle of projection

and g is the acceleration due to gravity.

In practice V is about $13\,\mathrm{ms}^{-1}$. Investigation of the distance thrown for various angles of release will show that $40°$ is nearer the optimum angle than $45°$; but above $35°$ the angle makes very little difference.

Consider various reasonable values of h. Other things being equal, what extra advantage does a tall man have?

The value of g differs in various parts of the world. At the equator it averages $9.78\,\mathrm{ms}^{-2}$ whereas at $60°$N at Leningrad it is $9.82\,\mathrm{ms}^{-2}$. What difference does this make? (It makes a difference of about $7\,\mathrm{cm}$.) Should anything be done about this when compiling world records?

Reference: G. H. G. Dyson, *The Mechanics of Athletics*, University of London Press, 1962.

Manufacture of bottles

Suppose you have 100 pounds of molten glass and you want to make 100 bottles each weighing one pound. In this molten glass are 100 small solid particles, called 'stones', distributed at random. If such a stone gets into the material of a bottle, that bottle is useless and must be discarded. How many bottles will be good? Using elementary ideas of probability the expected proportion of good bottles may be found by direct calculations as

$$\left(1 - \frac{1}{100}\right)^{100} \approx 0.366$$

but students who have not reached this theoretical level may organize a simulation using random numbers. By taking a sequence of 100 pairs of random numbers starting, for example, with

$$47, 82, 24, \ldots$$

we may allocate the first stone to bottle 47, etc. and investigate what happens. In this way a result may be obtained which is close to the theoretical distribution.

stones per bottle	0	1	2	3	> 3
bottles	36.6	37.0	18.4	6.1	1.9

Reference: A. Engel, *Teaching Probability in Intermediate Grades*, International Journal of Mathematical Education in Science and Technology 2, 243–294, 1971.

Designing a tin can

General results in algebra can often be studied in the first place by 'experimental' arithmetic. Consider the problem—the sum of two numbers is 12, what can their product be? Experiments with this, and with other fixed sums such as 20, 24 or 30, can provide a lot of practice at simple arithmetic and lead to the general conclusion that to maximize the product when the sum is fixed we must take two equal numbers.

A similar conclusion arises with three numbers. If the sum of three numbers is 24 the product is greatest when we choose 8, 8 and 8; and the same holds for other fixed totals. Accepting this as an experimental fact (an algebraic proof can come later) we may use it to solve a number of problems that are normally thought to require calculus. The total surface area of a cylindrical can, which has a base and a lid, is in the usual notation

$$A = 2\pi r^2 + 2\pi rh.$$

The volume is

$$V = \pi r^2 h.$$

The cost of the material for the can is roughly proportional to the surface area. How can we get the maximum volume for a given surface area; that is, how can we use the tinplate most efficiently? Consider the three quantities

$$2\pi r^2, \qquad \pi rh, \qquad \pi rh.$$

Their sum is A, which is fixed. Their product is a maximum when they are equal (by our agreed principle), and this product is

$$2\pi^3 r^4 h^2 = 2\pi V^2.$$

But making $2\pi V^2$ as big as possible also makes V as big as possible. To make $2\pi r^2$ and πrh equal means making $2r = h$, and this designs the most economical can. The diameter must be made equal to the height.

Note that the algebraic manipulation required to achieve this result is very small indeed; but this example is more often than not left until the sixth form in the belief that it requires calculus. Many similar examples are given in *Doing without calculus* by T. J. Fletcher in *Mathematical Gazette*, LV, 391, 4–17. Feb. 1971.

In science lessons at school a circular piece of filter paper is folded into a cone in a certain way, to use in a filter funnel. Does this method of folding produce a cone of maximum volume?

The ways to build a box

To build a rectangular box from rectangular wooden boards of non-zero thickness a carpenter must first decide which board overlaps which at each of the 12 edges. This suggests $2^{12} = 4096$ possibilities. However, not all 4096 are constructible; moreover, two possibilities which can be obtained from each other merely by rotating or reflecting the box ought not to be counted as different ways. In general, a rectangular parallelepiped can be built in 99 ways, and a cube in 18 ways.

In particular, the reader may like to prove for himself that there is no way of building a cubical box out of 6 congruent cuboidal pieces of wood (such as one can easily cut off a plank), and that a cuboidal box can be built from such congruent pieces of plank only if the lengths of the sides are suitably chosen.

The full analysis of these problems involves various elementary topological ideas, and some metrical notions as well. Some more advanced ideas such as direction fields on a sphere can also be employed.

Reference: E. N. Gilbert, *The ways to build a box*, Bell Telephone Laboratories (internal paper), Murray Hill, New Jersey, USA.

Shunting of railway wagons

The use of binary notation enables a number of decisions to be taken automatically. One such application is to the shunting of railway wagons. Suppose that the wagons are in a line as follows:

G I K D J C F A I E D A B H D

and that they have to be arranged in alphabetical order. We imagine that there is an engine to the left and a set of points with two sufficiently long sidings away to the right. What is a convenient method to get them sorted into the correct order, with as few shunting operations as possible? Note that the two wagons labelled 'I', for example, have ultimately to get to the same destination. They have to get to the appropriate place in the re-ordered sequence, but we do not mind which wagon 'I' comes first. The following is one algorithm.

We imagine that we are threading the wagons on a string, and we pass the string through them a number of times, each time picking up as many wagons as we can. A table should explain the process. The first pass is labelled 0, the second 1, etc., and those labels are then coded in binary form and attached to the various wagons.

	G	I	K	D	J	C	F	A	I	E	D	A	B	H	D
First 'pass'	.	.	.	.	.	.	.	0	.	.	.	0	0	.	.
Second 'pass'	.	.	.	.	.	1	.	.	.	.	1	.	.	.	1
Third 'pass'	.	.	.	2	.	.	.	.	.	2	.	.	.	.	.
Fourth 'pass'	.	.	.	.	.	.	3	.	.	.	.	.	.	.	.
Fifth 'pass'	4	.	.	.	.	.	.	.	.	.	.	.	.	4	.
Sixth 'pass'	.	5	.	.	.	.	.	.	5	.	.	.	.	.	.
Seventh 'pass'	.	.	.	.	6	.	.	.	.	.	.	.	.	.	.
Eighth 'pass'	.	.	7	.	.	.	.	.	.	.	.	.	.	.	.
First shunt	0	1	1	0	0	1	1	0	1	0	1	0	0	0	1
Second shunt	0	0	1	1	1	0	1	0	0	1	0	0	0	0	0
Third shunt	1	1	1	0	1	0	0	0	1	0	0	0	0	1	0

The binary labels are tabulated with the least significant digit at the top.

We have a scheme with labels involving three binary digits, and we will see that the shunting can be done with three operations. The first operation is to move the wagons to the sidings (imagined to the right), call the sidings 0 and 1, and set the points so that the wagons with a zero in the least significant place of their labels go to siding 0, and the wagons with a one go to siding 1. Now pull out the line of wagons from siding 0, and attach behind them the line from siding 1. This gives a train of wagons in order:

G D J A E A B H I K C F I D D.

Repeat the process using the middle digit of the labels as an instruction to set the points, getting eventually:

G A A B H I C I D D D J E K F.

A third operation, using the most significant digit, then puts the wagons in the required order.

With pupils it might be appropriate to introduce some simpler automatic process involving binary coding first; for example, the pupils can make the well-known set of number-guessing cards. A set of cards is constructed bearing the numbers

$$1, 3, 5, 7, 9, 11, 13, 15, \ldots$$

$$2, 3, 6, 7, 10, 11, 14, 15, \ldots$$

$$4, 5, 6, 7, 12, 13, 14, 15, \ldots \text{ etc.}$$

where the nth card carries the set of numbers whose binary representation has a digit 1 in the nth place, counting from the right (n cards sort 2^n numbers). The person doing the trick asks someone to think of a number and tell him which cards in the pack carry this number. The person doing the trick is then able to say which number is being thought about. (He simply adds up the first numbers on the cards named.)

It is possible also to code a set of n cards with binary numbers by using holes and slots along the edge. They can be sorted by using a knitting needle in these holes. What is essentially the shunting process just described will put the cards into numerical order.

Railway marshalling yards in practice often contain repeated bifurcations leading into 2^n sidings. Trucks are pushed over a hump and run into the various sidings according to the ways the points are set. The shunting method given above can, of course, be adapted to this type of yard.

Reference: R. Sprague, *Recreations in Mathematics*, Blackie, 1963.

Catalyst renewal

In a chemical plant a catalyst is often necessary to promote a chemical reaction. As the reaction goes on the efficiency of the catalyst diminishes, until eventually it has to be renewed. This involves shutting down the plant for a given interval of time τ. If we are given the productivity/time curve for the plant, starting with a fresh supply of the catalyst, when is the best time to shut down and renew?

In general, the decay curve for the productivity of the plant will be some complicated, empirically given curve. To simplify the problem we will assume initially that the decay curve is a straight line. There is no loss of generality if we choose units on the two scales so that the initial productivity is unity, and diminishes to zero in unit time.

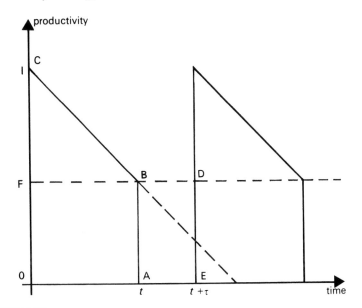

FIGURE 4.10

If the plant is shut down at time t and restarted at time $t+\tau$, when the catalyst has been replaced, the operation is described by the graph in Figure 4.10. The total production per cycle is given by the area OABC, which is

$$\tfrac{1}{2}t(1+1-t) = \tfrac{1}{2}t(2-t).$$

Since the cycle length is $t+\tau$, the average rate of production is

$$\frac{t(2-t)}{2(t+\tau)}.$$

There are various ways of continuing and it is instructive to compare them.

Method 1—Students who have learnt the early stages of the calculus may differentiate, and maximize this function of t.

Method 2—We may put $T = t+\tau$, the total cycle length, and say

$$\frac{t(2-t)}{2(t+\tau)} = \frac{(T-\tau)(2-T+\tau)}{2T} = \tfrac{1}{2}[2+2\tau-T-\tau(2+\tau)/T]$$

To minimize this it is necessary to minimize

$$T + \frac{\tau(2+\tau)}{T},$$

regarded as a function of T. But to do this is to minimize the sum of two terms whose product is constant. This may be done by using the converse of the principle used earlier (page 161). The sum is a minimum when the two terms are equal; this requires

$$T = \tau(2+\tau)/T$$

$$T = \sqrt{\tau(2+\tau)}$$

$$t = \sqrt{\tau(2+\tau)} - \tau$$

Method 3—We may argue much more in terms of the original situation. It is foolish to shut down the plant while the productivity is still above the best possible working average, also it is foolish to continue operating when the productivity has fallen below the best possible working average. This means that we must switch off when the productivity has fallen to its average value, which is indicated by the horizontal broken line. Hence:

$$\text{area BCF} = \text{area BAED}$$

$$\tfrac{1}{2}t^2 = \tau(1-t)$$

$$t^2 + 2t\tau = 2\tau$$

$$(t+\tau)^2 = 2\tau + \tau^2$$

$$t+\tau = \sqrt{\tau(2+\tau)}$$

$$t = \sqrt{\tau(2+\tau)} - \tau$$

Much can be learnt from this example. It is awkward to draw the graph of t as a function of τ, and easier to plot $\tau = t^2/[2(1-t)]$. Draw this graph and interpret it.

The third method of solution has further advantages over the other two. What should you do in a chemical works when the performance of the catalyst is not necessarily represented by a straight line, and when the performance of the catalyst may vary from batch to batch?

The Man with the Golden Gun

In most of the previous examples the mathematical manipulations involved are simple enough to aim to treat them in detail with a substantial number of pupils. But whilst mathematics is not a spectator sport there is something to be said also for the occasional descriptive example—an example where the teacher describes the mathematics only in general terms. This is worth doing if the situation is one which is sufficiently interesting for the pupils.

Those who saw the James Bond film, *The Man with the Golden Gun*, will have seen a chase in which a car jumped a river. The take-off being from a special ramp, the car not only jumped the river but did a 360° roll about a longitudinal axis as it went. This stunt was made possible only by extensive preparatory mathematical calculation and simulation on a computer.

A research department at Cornell University in the USA had been running a serious research programme on car accidents, and had devised a computer program which computed what happened when cars struck obstacles. It was possible also to produce pictures of what happens at various stages after the impact. The mathematics involved was largely classical dynamics, the study of which begins at school.

The computer program was adapted for the new problem, and a method found for launching the car into its spectacular jump. This involved the calculation of suitable launching angles and the appropriate launching speed. Thirty-three exploratory computer runs were needed for this. Without a computer this would mean quite a lot of wrecked cars, not to mention wrecked drivers; a further twelve computer runs were needed to design a suitable landing ramp.

This example shows the sensational side of applied mathematics, but perhaps the most satisfying aspect is that the original work was done as a contribution to accident prevention.

Reference: R. McHenry, *The astro spiral jump—an auto thrill show stunt designed via simulation*, Calspan Corporation, Buffalo, USA, 1973.

APPENDICES

Appendix A   The supply of teachers

In this book we have mostly been concerned with problems suited to the interests and technique of pupils in the 11–16 age range. Here we shall look at a problem of a different sort. The problem is at least a Believable one for teachers—how is the supply of teachers planned and what went wrong with the planning in the early 1970s.

The superficial features are well known from the newspapers—newly qualified teachers without jobs and the closure of colleges of education, some within a few years of their opening; there has been a lot of qualitative analysis in the press as well. (This publicity tends to make a problem suitable for modelling by informally providing the basic empirical background. Economic phenomena are similarly familiar to most people from reading the newspapers.)

It looks as though there has been some fairly incompetent planning of teacher supply. In these circumstances the applied mathematician starts to think how he would set about the problem, so let us look at it. There are many variants but the key ones must include the number of children and the number of teachers in the school system.

There are now two different approaches possible in building our model. We can either make an abstract algebraic model leaving the actual numbers until the end when we will have learnt more about exactly what numbers are needed, or alternatively, by finding out as many empirical facts as possible at the beginning, guide the modelling. In practice, because collecting facts takes time and energy, it is again usual to use an iterative compromise, starting with a few ideas, building a simple model, finding data to feed into that model, putting in some improvements, finding more data and so on.

So the simplest model of teacher supply might be to find the number of teachers either directly or from the number of pupils and the ratio of pupils to teachers.

$$\frac{\text{number of pupils}}{\text{ratio}} = \text{number of teachers} \qquad \frac{P}{R} = T$$

Then to find the necessary *supply* of teachers we can divide the number of teachers by the years of service between training and retirement, say $Y \approx 40$, to give the number required per year N.

$$N = \frac{\text{numbers of teachers in service}}{\text{years of service}} \qquad N = \frac{T}{Y}$$

This model will give a grossly unrealistic result for various reasons, but chiefly because many of those entering the teaching profession after initial training do not work as teachers continuously until retirement.

The model can be improved in a number of ways, step by step. We shall now collect together a number of improvements. The rate of teachers leaving the profession, and re-entering it, must be modelled, or more probably, measured. Similarly we must learn more about the ratio of teachers to pupils in different kinds of schools. It turns out that these ratios, which in 1976 were given as

$$\text{Primary } R = 24 \qquad \text{Secondary } R = 17$$

are very different from those that would be expected from average class sizes; this difference reflects the fact that not all teachers teach all the time, because they are involved in other supervisory duties or in the preparation of material. The other important factor which has to be taken into account is that the birth rate, which determines the school population in succeeding years, is not constant but has indeed been subject to large fluctuations over the last half century.

It is the fall in the birth rate below the predictions of the planners, due to social changes and perhaps to advances in contraceptive technique, which has been the main source of the imbalance of teacher supply. The time factor here is crucial. It takes five years for a new-born child to enter the school system and a further eleven to sixteen years before that year's birth rate ceases to affect the school population. It takes three years to train a teacher and probably about six years to build a college of education. In principle then it is possible to compensate for birth rate fluctuations in the teacher supply, but this may require fairly swift changes in plans of a type which the inertia of the political system makes quite difficult. The other factors we have mentioned of a social kind, which may quite suddenly reduce the movement of teachers out of teaching or of students into higher education, may demand even swifter responses.

Let us now try to make a simple but explicit mathematical model of the effect of the number of births $B(t)$ in year t on the pupil population $P(a,t)$ of age a in year t, as well as the corresponding teacher population $T(a,t)$ teaching age a, which depends on the staffing ratio $R(a)$. The equations

$$\left.\begin{aligned} R(a)T(a,t) &= P(a,t) \\ P(a,t) &= B(t-a)[1-D(a)] \end{aligned}\right\} \qquad \text{(A.1)}$$

show how the pupil and teacher populations depend on the birth rate in previous years. $D(a)$ is the drop-out rate—that fraction of the age group *not* in education. If these algebraic equations, connecting functions of two variables, seem rather forbidding, we can re-express them in less abstract form as a table giving, for each year, the number of pupils at various ages. We will make a rough attempt at this now. To do so we need a fair amount of information, some of which we can only guess; however, it is helpful to carry through a crude analysis of the problem in order to decide what information, among all the educational statistics published, we actually need to find in detail. In fact $R(a)$ and, particularly, $D(a)$ depend also on the time t, reflecting steady improvements in staffing (until a recent reversal) and changing fashions of staying on at school, as well as the raising of the school leaving age to 16 in 1974 (Figure A.1).

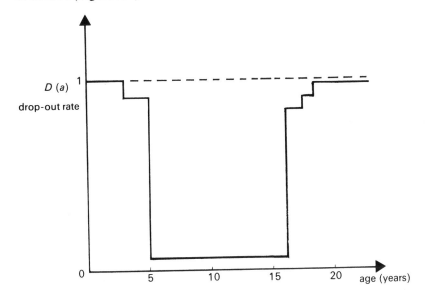

FIGURE A.1

The other essential piece of information is the birth rate function B. Figure A.2 gives the actual births up to 1977 and various projections for the future, based on different assumptions about fertility after the middle 1970s. The low variant is based on a continuation of the 1976 average of 1.8 live births per woman while, at the other extreme, the high variant takes the figure of 2.3; the long-term constant population level is 2.1.

From this data we can construct tables for $P(a,t)$. The DES report *School Population in the 1980s*, published in June 1978, gives projections for a total school population up to 1996 based on various assumptions about the number of children who will stay on at school after the age of 16—it is shown in Figure A.5. Notice how the peak reflects the peak in the birth rate in the late 1960s, with the swing delayed by about 10 years and somewhat reduced in extent by the smoothing out effect of summing over 11 successive years of birth rate.

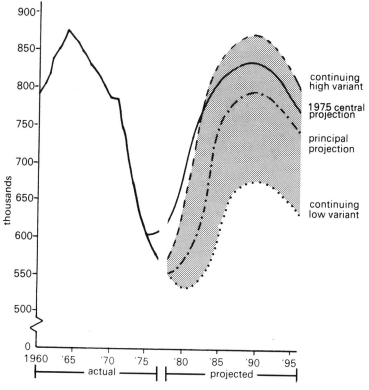

FIGURE A.2   Births in each year—England and Wales—1960 to 1977 and projections to 1996

Having decided how many teachers we need for each age group we return to the problem of the supply of new teachers $N(a,t)$. Algebraically,

$$N(a,t) = T(a,t) - T(a,t-1) + L(a,t) \qquad (A.2)$$

where L is the number of teachers leaving the profession during the year. Unfortunately L depends quite sensitively on external factors such as the economic climate which affects alternative employment. In summary,

$$N(a,t) = \frac{d}{dt}\left[\frac{B(t-a)(1-D(a,t))}{R(a,t)} \right] + L(a,t). \qquad (A.3)$$

Here, to make the equation more compact, we have treated t as a continuous variable; using

$$\frac{dX}{dt} \doteq X(t) - X(t-1). \qquad (A.4)$$

Exactly similar arguments can be carried through for the primary and secondary sectors separately and compared with Figure A.3.

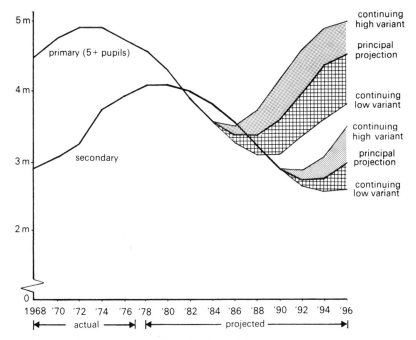

FIGURE A.3   Primary and secondary school population

Although (A.3) looks neater, we shall have to use (A.2) and (A.1) for numerical computation, since all our data will be discrete, coming at annual intervals; if we could proceed further using algebraic arguments, the continuous form would probably be more convenient. (This interplay between continuous and discrete mathematics is important, particularly at a more advanced level, but it can be seen and understood in the 11–16 range. An elementary illustration is provided by use of the calculator, which gives decimal answers when the actual situation demands a whole number; as with the number of Mars bars on page 37.)

The realistic graphs can only be handled numerically but we can get a clearer insight more simply by studying a simplified graphical model. We could usefully approximate the oscillations in the birth rate, either by the sine waves or by straight-line zig-zags; the former would demand that we use algebra so we shall choose the latter. We shall assume that the birth rate B averages 0.7M per year but oscillates by ± 0.2M (about 30% as found) with a 20 year period; for further simplicity let us assume that school lasts 10 years from age 5 to 15 for all pupils. Figure A.4 shows the birth rate B, the total school population $P = 20T$, and the number of new teachers N required assuming that 17 000 teachers ($\simeq 5\%$) leave teaching each year. You may notice the $\pm 30\%$ swing in the birth rate B is smoothed by summing the 10 age groups in school to a $\pm 15\%$ swing in the school population P, but that the effect on the demand for new teachers N swings dramatically and sharply by $\pm 60\%$ as T swings from rising to falling roles (mathematically the differentiation in (A.3) acting on the zig-zag produces a square wave illustrating the more general points that differentiation sharpens change while integration, as in moving from B to T, smooths out change). Thus the variation of N is a factor of about 4 from peak to trough. In working out the form of T from B, it is simplest to note the 'average' height of the B curve over the interval which stretches 5 years either side of the 'age 10' birth date (10 years before the time in question); in calculating N, notice that T rises, then falls, by 100 000 over 10 years, i.e. by 10 000 per year.

Although there are not yet any adequate models of all the factors in the teacher supply situation, we can use what we have developed to make explicit numerical predictions as the basis of a rough comparison with data on teacher training. Even from this simple model, clear implications for colleges of education and their staff will emerge. Assuming slow variation of the other factors, the peak in the school population shown in Figure A.5 implies an increase of about 8 000 teachers per year on the rising slope of the early 1970s and a similar number *fewer* in the early 1980s; given that teacher training lasts three years, this amounts to a reduction of about 50 000 in the number of teacher training places, with a corresponding reduction in the staff of the colleges.

However, these sharp reductions have already taken place. What other factors, not included in our model, may account for this? There are at least three that are of importance. Firstly, in planning teacher supply in the late 1960s it was assumed that the slow fall in the birth rate was a temporary feature and that, in any case, increased prosperity would

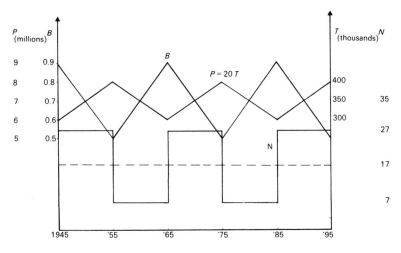

FIGURE A.4

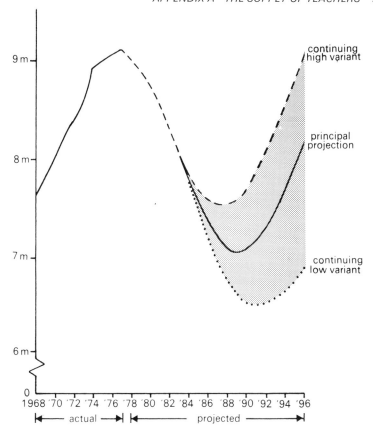

FIGURE A.5 Total school population

allow for improved staffing ratios; thus more teachers were trained over this period and more new teacher training places were created. Secondly, the very sharp drop in birth rate from 1972 onwards coincided with a period of severe economic difficulty in which, far from improving staffing ratios, the teaching force was actually cut, even though the school population was still increasing slightly. Thirdly, the economic situation also sharply reduced the movement of teachers into other sorts of employment—from about 12 per cent (of the total of about 400 000 teachers) each year to about 6 per cent; this factor alone would correspond to a reduction of 60 000 teacher training places.

In fact, it was decided to reduce the total number of teacher-training places in England and Wales from its peak of about 120 000 to 45 000 by 1981 with 10 000 of these to be used for in-service training of teachers. This amounts to a catastrophic change in one sector of the British educational system.

Appendix B A teaching unit on kinematics*

The following detailed teaching unit on the kinematics of traffic shows how practical problems can be tackled in a straightforward way, while still placing *some* demands on the students' modelling skills. The same approach could be used with many of the problems in this book.

Distance

The following table gives distances between towns in kilometres (km) above the diagonal, and in miles below the diagonal.

Distances	Bi	Br	Le	Lo	Ma	No	
Birmingham		140	175	177	?	85	
Bristol	87		312	187	?	220	
Leeds	109	194		306	?	108	km
London	110	116	190		?	198	
Manchester	79	159	40	184		?	
Nottingham	53	137	67	123	70		
				miles			

Table 1

1 Sketch a map of the towns in the table, with lines (curves if necessary) joining the towns. Write in the distances in miles and kilometres where given.

2 The distance from Birmingham to Nottingham is given as 53 miles or 85 kilometres. Calculate the number of kilometres in one mile using this pair of distances. Using as many more pairs of distances as you can, find the number of kilometres in one mile. Explain the method you used.

From the spread of your answers how accurate do you think this number is? Calculate the distances to Manchester in kilometres.

3 The usual route from Manchester to London passes close to Birmingham. How big a detour (how many *extra* kilometres) is required to go through Birmingham? Find two other routes which pass close to a third town and find how big a detour is needed in each case.

---

* This section was written with Frank Knowles.

Speed

Speed is a measure of the rate of movement, or rate of change of position. At a constant speed the distance travelled in any two equal time intervals t will be the same. (For example you will travel 200 km in 2 hours at $100 \, \text{km h}^{-1}$ on the M6 and then another 200 km in the next 2 hours after you branch off onto the M5 at Birmingham if you keep to $100 \, \text{km h}^{-1}$ on both sections of the motorway.) The speed is defined as

$$v = \frac{d\text{-}step}{t\text{-}step}$$

that is

$$\frac{\text{distance travelled}}{\text{time taken}}.$$

We can show this graphically by plotting the position of the car (or train or other object) at a selection of times t during its journey. This gives v as a measure of the slope of the curve. Since we have assumed v is constant, the curve has constant slope and is a straight line. The gradient of the line is v.

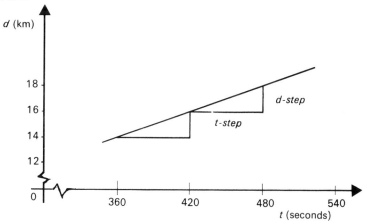

FIGURE B.1

We shall also look at variable speed situations when $d\text{-}step/t\text{-}step = \bar{v}$ is called the average speed during the time interval. We find that $\bar{v}$ varies depending on which time interval we take, and how long the interval is. What is happening in the graph in Figure B.2?

On a long car journey the average speed = distance travelled/time taken, $\bar{v} = d/t$. The car speedometer shows the instantaneous *speed* which may be thought of as the average speed over a very short time interval.

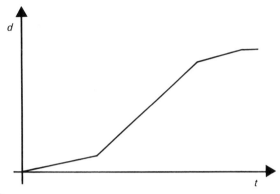

FIGURE B.2

1 A lorry on a motorway has the distance–time graph of Figure B.3. Find its speed. Rearrange the formula $\bar{v} = d/t$ to find (a) an expression for the time t taken to travel a distance d at average speed $\bar{v}$, and (b) the distance travelled in time t at average speed $\bar{v}$.

   This gives a total of three versions of the same basic formula. Check all three formulas on the lorry driving down the motorway.

2 A sprinter A runs 100 metres in 10.8 seconds. Calculate his average speed.

   A second runner B runs 100 metres in 10.6 seconds; what is his average speed? How far ahead of the first runner is he at the finish, assuming that they start together?

   Assuming they run at constant speed, draw a (straight line) graph of distance against time for each runner. (Use the same axes for both graphs.)

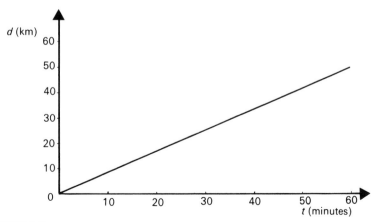

FIGURE B.3

3 A bus takes 18 minutes to go four miles from the city centre to an out-of-town bus stop. Find its average speed. When it is moving, it averages 20 mph. Find how long it spends stationary at bus stops, etc.

4 A bicycle in heavy traffic maintains a constant speed of 15 mph. Cars travel at 30 mph or are stationary. What fraction of the time must the car be moving if it is to take the same total time as the bicycle? (Try this for a cyclist riding for 1 hour, 2 hours, and t hours.)

5 A long-distance transport company has target journey times for its lorries between towns. The following table gives these times in hours.

	Br	Le	Lo
Br			
Le	4.8		
Lo	2.9	4.7	
M	3.9	1.0	4.6

Table 2

Assuming a constant speed of travel, about what speed does the company expect its drivers to keep to?

Clearly traffic cannot always travel at constant speed. What is the meaning of the assumed speed the lorries have to keep to?

If the assumed speed were increased to $80 \, \text{km h}^{-1}$, how would the times be changed? For this speed recalculate Table 2, extending it to all the towns in Table 1. Give the values to the nearest 6 mins ($= 0.1$ hours).

6 A commuter lives outside a town. On his journey to work he goes half the distance (8 km) at $60 \, \text{km h}^{-1}$ and the other half in heavy traffic at an average of $20 \, \text{km h}^{-1}$. Calculate the times t_1 and t_2 he spends on each half. Then calculate his average speed.
Why is this not $40 \, \text{km h}^{-1}$?

$$\left(\frac{60 + 20}{2} = 40 \right)$$

Draw a position–time graph to illustrate this situation.

Some other words you may hear used sometimes include:

displacement = distance moved = change of position

velocity = speed = rate of change of position

(Strictly, velocity means speed in a particular direction, so that when going round a corner at a constant *speed* the *velocity* changes.)

7 An automatic forklift truck in a warehouse repeatedly travels 4 m forwards in 12 seconds, waits 8 seconds while it picks up goods, goes 7 m backwards in 35 seconds, waits 8 seconds to put them down, and finally goes back 3 m to its starting point in 9 seconds.
Draw a distance–time graph for the travels of this forklift truck.

Acceleration

Figure B.4 shows the speed–time graph for a car leaving traffic lights at $t = 0$. Describe what is happening.
  – The car accelerates uniformly for 5 seconds from rest to $10 \, \mathrm{m\,s^{-1}}$, then it goes at constant speed.

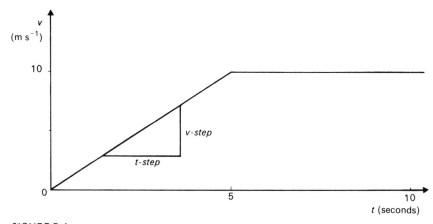

FIGURE B.4

Acceleration is a measure of the rate of change of speed with time. It is normally measured in units of metres per second per second, written $\mathrm{m/s^2}$ or $\mathrm{m\,s^{-2}}$. As in speed itself, we distinguish the average acceleration over a time interval

$$\bar{a} = \frac{v\text{-}step}{t\text{-}step}$$

and the instantaneous acceleration at a given time, which may be thought of as the average as the time interval around that time becomes shorter and shorter. If the acceleration, a, is a *constant* over a time interval, all averages and instantaneous accelerations within this interval are equal to a. What is the average acceleration from $t = 0$ to $t = 5$ in Figure B.4?

$$\bar{a} = \frac{10 - 0}{5 - 0} = 2 \, \mathrm{m\,s^{-2}}$$

What is the average acceleration from $t = 5$ to $t = 10$ seconds?

$$\bar{a} = \frac{10-10}{10-5} = 0\,\mathrm{m\,s^{-2}}$$

So the average acceleration is $0\,\mathrm{m\,s^{-2}}$. In both these cases the acceleration is also constant throughout the intervals. What is the average acceleration from $t = 0$ to $t = 10$ seconds?

$$\bar{a} = \frac{10-0}{10-0} = 1\,\mathrm{m\,s^{-2}}$$

So the average acceleration is $1\,\mathrm{m\,s^{-2}}$. This is less than the average acceleration of the first 5 seconds because it assumes that the acceleration of $1\,\mathrm{m\,s^{-2}}$ was constant over *all* 10 seconds, when in fact it changed at $t = 5$.

In all this you may have noticed that,

acceleration, speed and time

have the same relationship as

speed, distance and time

in our previous discussions. (They are not the same things, of course.) The relationships may be summarized by

$$\bar{a} = \frac{v\text{-}step}{t\text{-}step} \qquad\qquad \bar{v} = \frac{d\text{-}step}{t\text{-}step}$$

1 Sketch the speed–time graphs, like Figure B.4, for:
    a your bus from home to the town centre
    b an aeroplane flight from London to Paris
    c a racing car on a track whose shape you draw first.
    d a parachutist falling from an aeroplane
    e an athlete running 100 metres, and 1500 metres
    f the vertical speed of a high-jumper.
2 What are the accelerations in the various stages of the motion in the three graphs in Figure B.5?
3 Criticize and improve Figure B.4 as a realistic picture of a car accelerating from traffic lights.

What about distance?

So far we have talked mostly about time, speed and acceleration. We are still interested in the distance travelled. We saw that the distance travelled in time t at constant speed v is

$$d = v \times t.$$

It is given by the area of the speed–time graph, Figure B.6.

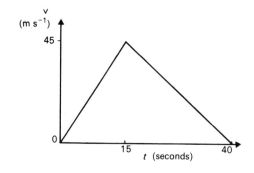

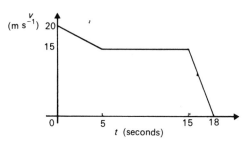

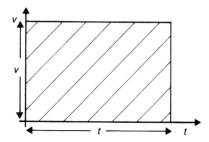

FIGURE B.5

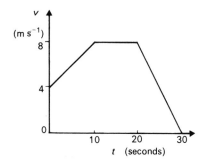

FIGURE B.6

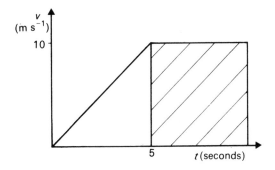

FIGURE B.7

If the speed is not constant, we can extend this idea by looking at the area of different speed–time graphs. In Figure B.7 the distance travelled between 5 seconds and 10 seconds is $10 \times 5 = 50\,\mathrm{m}$. The distance travelled between 0 and 5 seconds is found by the same method of area. Here it is the area of a triangle, and is given by

$$\tfrac{1}{2}(\text{height} \times \text{base}) = \tfrac{1}{2} \times 10 \times 5 = 25\,\mathrm{m}$$

so the total distance is $75\,\mathrm{m}$. If the speed gain is $v\,\mathrm{m\,s^{-2}}$ in time t and the acceleration is constant, then the area, shown in Figure B.8, is $\tfrac{1}{2}vt$, and this represents the distance travelled.

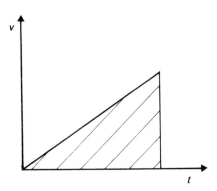

FIGURE B.8

So now we have for *constant acceleration* starting from rest:

$$d = \tfrac{1}{2}vt$$

where v is the final speed. But $v = at$, so

$$d = \tfrac{1}{2}\,at \times t = \tfrac{1}{2}at^2.$$

Notice that the distance travelled goes like the square of the time. This is because speed and time are both proportional to time and are multiplied together, to make speed go like the square of time.

Now let us plot the distance moved by the car in Figure B.4; note the quadratic shape from 0 to 5 seconds (while the car is accelerating) and the linear distance/time relationship when the speed is constant after 5 seconds, as in the previous section. The gradient of this curve at any point is the speed at that time (Figure B.9).

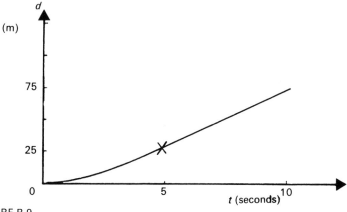

FIGURE B.9

1 Find the distances travelled in the various stages of the graphs of Figure B.5.
2 Find the total distance travelled by a car that gets up to $30\,\mathrm{m\,s}^{-1}$ in 15 seconds and then travels for 30 seconds at $30\,\mathrm{m\,s}^{-1}$, and then stops with constant acceleration in 20 seconds. (Draw a sketch graph first.)
3 The driver of a fast car accelerates steadily from 0 to $30\,\mathrm{m\,s}^{-1}$ in 10 s, and remains at that speed. Starting at the same time a slower car accelerates steadily from 0 to $30\,\mathrm{m\,s}^{-1}$ in 15 s, and remains at $30\,\mathrm{m\,s}^{-1}$. Find the distance travelled by the cars when accelerating, and so find how far ahead the fast car is. Will this distance remain the same?

What about starting not from rest (constant acceleration)?

Let's look at the situation described in Figure B.10.

If the speed is u at $t = 0$ and v at time t, then the acceleration a is given by

$$a = \frac{v - u}{t}$$

so $at = v - u$ and

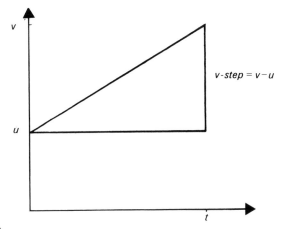

FIGURE B.10

$$v = u + at$$

Using the same idea as before, that the area under the graph represents the distance, we need to find the two shaded areas in Figure B.11.

These areas are indicated by the arrows in the figure and we see that the total distance travelled is

$$d = ut + \tfrac{1}{2}at^2$$

These are the basic equations for constant acceleration.

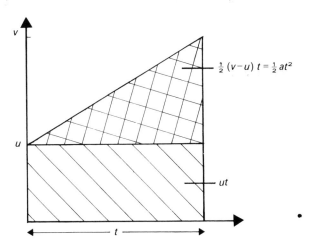

FIGURE B.11

1 The force of gravity alone provides a constant downward acceleration $g = 10\,\mathrm{m\,s^{-2}}$. If a body falls from rest, find how fast it is moving and how far it has fallen from rest after 1, 2, 10, 20, 100, 200, 1000 seconds. How does this differ from what happens to a parachutist?

2 A car driver makes an emergency stop from a speed of $30\,\mathrm{m\,s^{-1}}$. For one second after the emergency the car continues at the same speed while the driver reacts to the situation; from then on the brakes provide a constant negative acceleration $a = -5\,\mathrm{m\,s^{-2}}$ until the car stops. Sketch the speed–time graph for the motion of the car and find out how far it travels after the emergency before it stops.

Calculate the stopping distance for initial speeds of 5, 10, 15, 20, 25 and $50\,\mathrm{m\,s^{-1}}$, assuming the same reaction time and braking acceleration ($10\,\mathrm{m\,s^{-1}}$ is roughly 20 mph).

3 When testing new cars, motoring magazines usually give the time for the car to accelerate to a given speed. Look up some of these figures and calculate the average accelerations. How do they compare with the braking acceleration given in the previous question? How would you expect the actual instantaneous acceleration to depend on the speed of the car?

4 When a car travelling at $30\,\mathrm{m\,s^{-1}}$ runs into a brick wall it stops in about 1 metre. Estimate roughly how long this takes and thus the average acceleration during this period. Is a model with uniform acceleration a sensible one? If not, why not? How might one expect it to depend on time or current speed?

Project—Timetabling the railway line between Redcliffe and Camden

Major cargo traffic
The area exports iron ore through Camden, imports steel for industry in Ashburn, and exports fruit and vegetables through Camden via South Amherst.

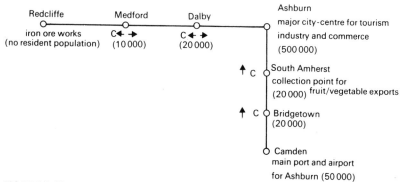

FIGURE B.12

Major passenger traffic

The line carries mostly commuter, tourist and business traffic between Ashburn and Camden. There are no long distance lorry routes, and most cargo and passenger traffic is by rail.

Requirements and restrictions

1 Six iron ore trains per day to Camden are needed (and return to Redcliffe). These six trains must travel in the daytime, as there is no storage space or night-time loading at Camden.

2 Passenger trains are needed to suit usual commuting and business and tourist traffic. (C indicates commuting to work in the direction of the arrows.)

3 Overtaking of trains is allowed only at Ashburn and Medford.

4 Trains may be of four types: fast passenger, slow passenger,
goods/parcels, iron ore in bulk

Timings and distances between each town in each direction are given below. N.B. You will need to make other assumptions within the situation given.

Timings (in minutes) and distances (in km)

	Fast passenger	Slow passenger	Goods and Parcels	Iron ore	Between stations	From Camden
	Times (minutes) and *train type*				*Distances* (km)	
Redcliffe						180
	15 / 18	21 / 21	22 / 21	24 / 30	32	
Medford						148
	18 / 18	21 / 21	21 / 24	24 / 33	36	
Dalby						112
	15 / 15	18 / 18	18 / 18	24 / 30	28	
Ashburn						84
	24 / 24	24 / 24	31 / 30	33 / 42	40	
South Amherst						44
	15 / 12	18 / 15	21 / 18	24 / 21	26	
Bridgetown						18
	12 / 9	15 / 12	15 / 12	15 / 15	18	
Camden						0

Key

Timings: away from Camden/towards Camden.

For through timings and distances add individual timings and distances.

What you have to produce:

1 A statement of assumptions you have made
2 A weekday passenger timetable for *one direction only* for one day
3 A train graph for *all* trains in *one direction only* for one weekday
4 A conclusion, which need not be long.

Advice

1 Use 3 or 4 pieces of graph paper sideways.
2 Use a scale of 2 large squares (4 cm) to 1 hour or 1 small square (2 mm) to 3 minutes or 1 large square (2 cm) to 10 km.
3 Colour coding is useful.

INDEX